# A Systems Approach to Environmental Management

## It's not easy being Green

Related titles published by Dunedin Academic Press:

*Sustaining Coastal Zone Systems* (2011)
Edited by Paul Tett, Audun Sandberg and Anne Mette
ISBN: 978-1-906716-27-1

*Introducing Oceanography* (2012)
David N. Bowers and David C. Bowers
ISBN: 978-1-78046-001-7

For further details of these and Dunedin's other titles see
www.dunedinacademicpress.co.uk

# It's not easy being Green

# A Systems Approach to Environmental Management

**Tim O'Higgins**

*Scottish Association for Marine Science, Oban*

*with a chapter by*

**Mohammed Saif Al-Kalbani**

*Centre for Mountain Studies, Perth College*
*University of the Highlands and Islands / University of Aberdeen*

EDINBURGH ◆ LONDON

*Published by Dunedin Academic Press Ltd*
*Head Office:* Hudson House, 8 Albany Street, Edinburgh EH1 3QB
*London Office:* 352 Cromwell Tower, Barbican, London EC2Y 8NB

ISBNs
978-1-78046-055-0 (Hardback)
978-1-78046-026-0 (Paperback)
978-1-78046-533-3 (ePub)
978-1-78046-534-0 (Kindle)

© Dunedin Academic Press 2015

The right of Tim O'Higgins and
Mohammed Saif Al-Kalbani to be
identified as the authors of their parts of
this work have been asserted by them in
accordance with sections 77 and 78 of the
Copyright, Designs and Patents Act 1988

## bitlit

A **free** eBook edition is available
with the purchase of this print book.

CLEARLY PRINT YOUR NAME ABOVE IN UPPER CASE

**Instructions to claim your free eBook edition:**
1. Download the BitLit app for Android or iOS
2. Write your name in **UPPER CASE** on the line
3. Use the BitLit app to submit a photo
4. Download your eBook to any device

*British Library Cataloguing in Publication data*
A catalogue record for this book is available from the British Library

Typeset by Makar Publishing Production, Edinburgh
Printed in Poland by Hussar Books

# Contents

# List of illustrations

# Acknowledgements

The substance of this book stems from KnowSeas, an international research project, which I coordinated under leadership of the late Professor Laurence Mee. The project examined the environmental management of Europe's seas (which explains the marine 'flavour' of the book) but the ideas developed can be applied anywhere. I owe a great debt of gratitude to Laurence Mee for encouraging me to develop this book and giving me the time to do it. I am sorry that he is not here to read the final product. Thanks also to all my friends and colleagues at the Scottish Association for Marine Science, at the University of the Highlands and Islands, and all those around Europe who worked so hard on the KnowSeas project.

I would not have written this book without Dr Thom Nickell, with whom I shared many enjoyable conversations and whose strength of conviction inspired me to write down some of our ideas. Thanks also to my wife, Linda, for supporting me in this work and my daughters for providing critical appraisal of the graphics.

I hope that the reader will find this book interesting and that some of the ideas here will help a new generation of environmentalists to realize that managing societies and managing the environment are pieces of the same puzzle.

*The research leading to these results has received funding from the European Community's Seventh Framework Programme [FP7/2007-2013] under grant agreement number 226675.*

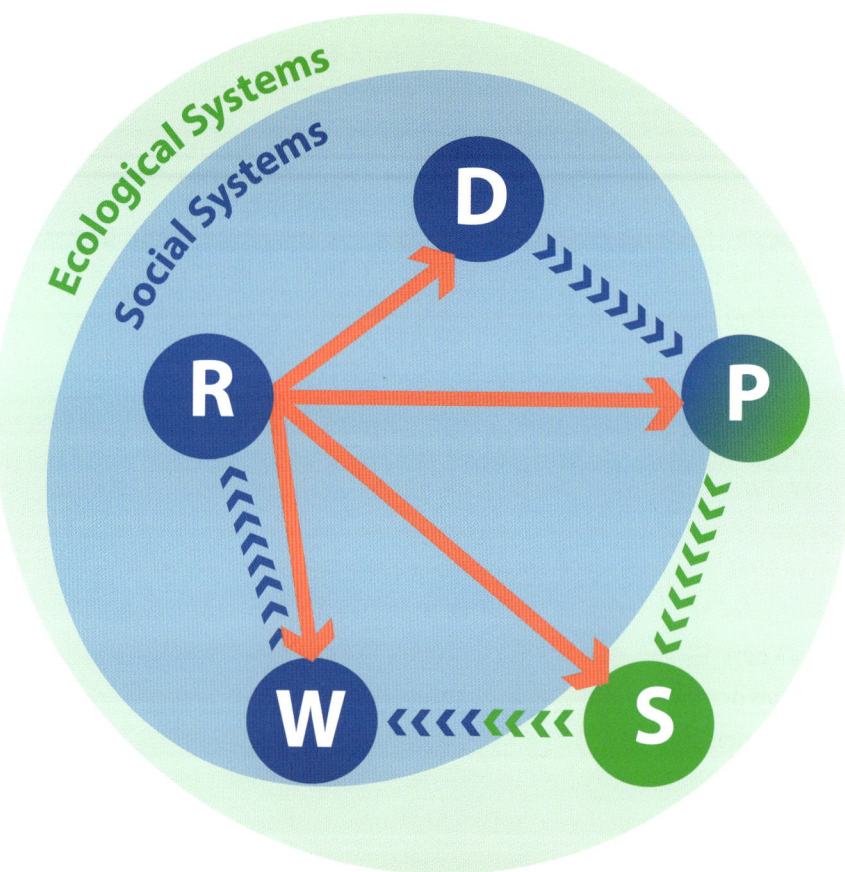

# Introduction[1]

All human activity, every meal, every breath, has environmental consequences. Understanding the consequences of our actions is vital to understanding (and correcting) our relationship with nature. In an era of globalization, where supply chains snake across borders and goods and services may be produced or consumed practically anywhere on the planet, tracing the impacts of individual actions on the Earth's ecosystems is becoming increasingly difficult. Managing them is even more complex. Some of our activities can be changed and adapted; food can be obtained from alternative sources or by alternative techniques; others, like breathing (which contributes greenhouse gases to the atmosphere) are fundamental to our biology.

We are inundated with examples of ecological catastrophes. From climate change to pollution of air and water, depletion of fish stocks or destruction of habitats, losses of biodiversity and extinctions of animals and plants – examples where natural resource management needs to be improved are everywhere. But environmental problems are almost invariably complex. Societies do not set out to cause damage to their surroundings, and economic development is conducted with the goal of improving human welfare. The environmental problems caused by development are usually unintentional, unconsidered, unavoidable or too costly to curtail. Ecological problems are very often economic **externalities**, that is, their costs are not reflected in the prices of our economic activities. For example, the cost of a river polluted by cattle farming is not included in the price of a hamburger.

Concentrations of carbon dioxide and methane in the atmosphere are elevated because fossil fuels have been used to drive economic growth. Societies depend on fossil fuels to meet fundamental needs, in the transport of

1 The chapters of this book sequentially cover different aspects of the DPSWR framework. At the start of each chapter you will find a diagram of the DPSWR showing which elements are covered in the following chapter (for explanation *see* Fig. I.2).

food, the provision of heat and to travel from place to place. Rainforests are cleared for agriculture to produce more food or for precious hard-woods (Geist and Lambin, 2002). Fish stocks collapse to meet demand for food. Each of these archetypal environmental problems involves the generation of human welfare at the cost of the environment. At the heart of most, if not all, such management problems lie connections between improved human welfare and their unintended and undesirable consequences. But ecological degradation has costs too. Increased carbon dioxide in the atmosphere has been linked to increases in wildfires, drought and flooding (Scholze et al., 2006). The clearance of rainforests results in biodiversity loss and species extinctions (Bradshaw et al., 2009) as well as losses in non-timber forest products (Yaron, 2001; Balmford et al., 2002). The collapse of fisheries results in losses to human well-being through loss of food security, through unemployment, social decline and migration (Coulthard et al., 2011; Hamilton and Butler, 2001).

Recognizing the trade-offs between the benefits of economic development and the costs of environmental degradation, making them explicit and incorporating them in development decisions is essential to good management. Since human society is utterly dependent on the environment for survival, effective environmental management is critical to human welfare on a global scale.

Natural resources are often **common pool resources**. Ownership of these resources is collective, but the benefits of exploiting them may flow only to specific groups while the costs of environmental damage are borne by others. As a result, environmental problems are often contentious and depend on differing values placed on the elements of the world around us. When setting environmental targets, different groups of users with different sets of values, cultures, laws, histories and aspirations must also be considered and each may have differing objectives. The right course of action or the best plan for management depends on individual perspectives. These considerations are at the heart of the increasingly popular, but difficult to define, concepts of Ecosystem Based Management and the **Ecosystem Approach** to management. Paradoxically, they are founded on the inclusion of humans (and the social systems they produce) as part of the ecosystem itself.

The focus of this book is a systematic technique which can be used to break down the complexity of environmental management decisions so that social and ecological aspects are considered in tandem. Whatever the particular management objectives, the method enables development of a full picture of the trade-offs inherent in a particular decision.

Before describing the theoretical framework, it may first be useful to present an illustrative example of a typically complex, contentious and intractable management dilemma. The example of the management of the Columbia River Salmon Fishery illustrates well many aspects of the complexity involved in natural resource management. The following section provides an overview of some ecological and social challenges posed by the management of the Columbia River, a 'perfect storm' of complexity summarized in Figure I.1.

## The Columbia River Salmon Conundrum

The Columbia River is the second largest river in the United States. Rising in the Cascade Mountain Range its watershed covers an area of 668,000km$^2$ and straddles the international boundaries of the US and Canada and state boundaries of Oregon, Washington, Idaho, Montana and Nevada. The river flows westward into the Pacific Ocean forming the borders between two US states, Oregon and Washington.

Salmon are iconic in the Pacific Northwest Region of North America as symbols both of its history and of its natural abundance. Five major species of salmonid live in the Columbia River: Chinook, Coho, Sockeye, Steelhead and Chum. Salmon populations supported major subsistence fisheries for about 9000 years before the arrival of European settlers (Butler and O'Connor, 2004). These fisheries were vital to survival of the local tribes and they appear to have been relatively stable over time. The salmon supported the development of relatively dense human populations estimated at 50,000 inhabitants and the tribal harvest to meet the needs of these populations may have been in excess of 8000 tonnes annually (Craig and Hacker, 1940).

The management of the salmon was culturally engrained in the practices of the native peoples (Campbell and Butler, 2010). Following the arrival of European settlers, Native American populations declined rapidly in numbers due, in great part, to the ravages of new diseases including smallpox, malaria, and measles carried by the newcomers. In the latter part of the 1800s, around the same time as the decline in numbers of the indigenous peoples, a larger

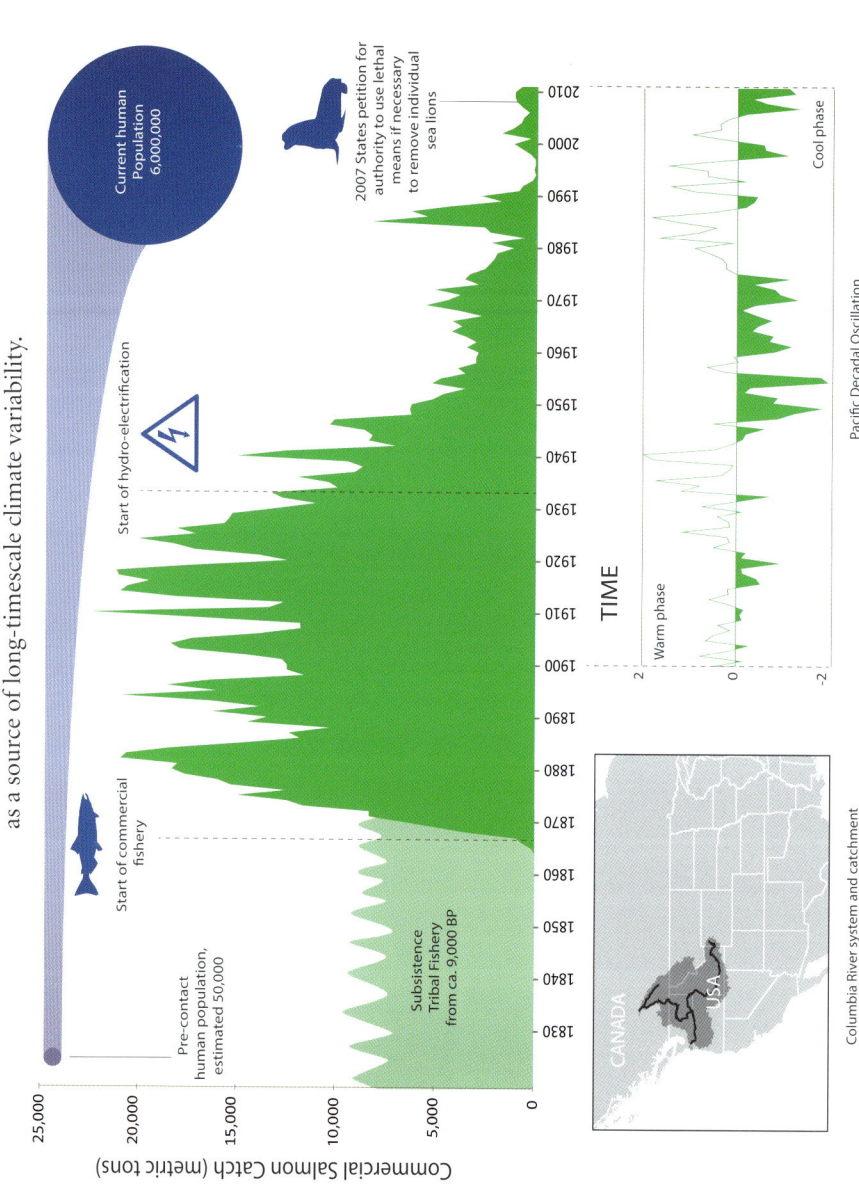

**Figure I.1** Timeline of major events in the Columbia River salmon fishery showing multiple competing Drivers and the Pacific Decadal Oscillation as a source of long-timescale climate variability.

(but comparable) scale commercial fishery began to develop. Commercial catches fluctuated periodically but remained above the estimated pre-commercial levels into the early part of the twentieth century.

The next major development in the timeline of Columbia salmon history was the construction of a series of hydroelectric dams in the basin. The dam construction in the 1930s paved the way for electrification, modernization and development of the Pacific Northwestern region. The programme of electrification was part of 'The New Deal', a series of US federal economic policies put in place to counteract the devastating effects of the Great Depression.

The dams (the largest being the Bonneville dam, which began construction in 1934) were built for rural electrification and allowed for rapid growth of human populations in the Columbia River catchment (now approximately 6 million). But they also had an unintentional consequence. They obstructed the passage of fish, and the lakes they generated resulted in the loss of fish habitat. For example, the Grand Coulee dam completed in 1941 effectively blocked salmon passage to 1800km of the Columbia River. Following electrification commercial salmon catches began to decline. By the late 1930s mean annual catches were half their maximum volumes. By the mid-1990s the lowest ever commercial catches were recorded (Figure I.1).

The decline in salmon catches has prompted major programmes for restoration and remediation. These efforts have focused on hatchery programmes; estuarine and tributary habitat restoration projects; adaptation of the dam structures to improve salmon passage, including the construction of spillways; harvest controls, restrictions on the take of salmon and more recently, predator controls (Naiman et al., 2012). There have been extraordinary and imaginative efforts and ideas, including the transport of young fish by barge from the upper portions of the river past the hydroelectric dams and bounties placed on the pike minnow (a predator of juvenile salmon). In recent years, predation of returning salmon by California sea lions has been a problem. State governments have applied for federal licences to use lethal force where necessary to remove sea lions normally protected under the US Federal Marine Mammal Protection Act (1972) and this has led to a legal battle with animal welfare advocates. Annual costs to improve salmon runs exceed US$600 million per year including opportunity costs of temporarily suspending electrical generation to allow salmon passage (Bonneville Power Administration, 2012).

While the science of remediation is continuing to grow and develop, these efforts must be set against the backdrop of an ever-changing system with **exogenous** (uncontrollable, external) influences. The entire region is affected by the Pacific Decadal Oscillation (PDO) which results in natural fluctuations in marine water temperatures, in turn affecting the success of salmon runs on an annual basis. This causes variations in the numbers of fish available for a sustainable commercial catch (Mantua et al., 1997; Mantua and Hare, 2002). The fluctuations of the PDO also confound attempts to separate out natural and man-made variability in salmon runs, and the relative success of different restoration activities. Despite all these efforts the fact remains that in the last 20 years commercial catches have never been more than 10% of their historical maximum. Tribal and non-tribal commercial fisheries as well as recreational fisheries have all suffered the effects of the decline. Currently 13 stocks of salmon and steelhead are listed as threatened or endangered (Ford, 2011).

Enormous thought and effort have gone into the management of the Columbia River Salmon. It is impossible here to give more than a flavour of the problem. Lackey (2000) gives an excellent overview of the salmon management challenge for the Pacific Northwest.

The point of the example is to illustrate some features of the complexity of environmental management problems in general. There are many different users of the system with different and potentially conflicting objectives. Different activities result in a variety of stressors to the system which cause changes to the system itself. The changes to the river have had unforeseen and undesirable consequences. There is no clear solution.

The solutions to problems such as the Columbia River salmon case described above cannot be answered by any single discipline. Ecologists can supply information on the state of the salmon stocks and levels of catch that might bring about improvements in salmon numbers, or the relative importance of the Columbia estuary California sea lions to the west coast population. Economists can provide information on the costs of loss of the commercial fishery (Radtke, 1993) and costs of restoration efforts (Landry, 2003) or the benefits generated by the electrification of the region. Historians and anthropologists can analyse the cultural significance of the salmon fishery to the native people and the early European settlers (Campbell and Butler, 2010; Butler and O'Connor, 2004). Ultimately however, political and legal decisions govern what environmental management actions are taken, and these decisions involve trade-offs, judgements, choices and values which may or may not be explicitly considered.

Determining the right course of action requires consideration of social, economic, cultural and ecological factors. Under such complex conditions systematic characterization of particular problems can enable the decision maker to articulate the trade-offs and develop a coherent and adaptive approach to management. Recognizing the complexity of problems and identifying the many interacting disciplines required to understand different parts of these problems is essential to good management. The aim of this book is to introduce and explain a problem structuring framework to facilitate such a holistic and systematic approach to environmental management.

## The Driver Pressure State Welfare Response Framework (DPSWR)

Figure I.2 illustrates the **Driver Pressure State Welfare Response (DPSWR)** conceptual model and defines each of the information categories within it. The conceptual model is presented not as a proscriptive all-encompassing model but a way of thinking about problems which can help to simplify and clarify the interrelationships between different components of the system. Cooper (2013) provides a thorough, meticulous and detailed account of the development of the framework, its various adaptations and modifications to its present form.

The elements in green represent the ecological system, while the blue elements make up the social system. Note how the social system is contained within the ecological system. The remainder of this chapter illustrates briefly how the DPSWR can be used to categorize the elements of the Columbia River management problem described above in order to understand each element of the problem separately in the context of the system as a whole.

**Drivers** are considered to be the economic and social forces that result from government policies, markets and the activities of private industry. In the Columbia River example there are many competing Drivers of the ecosystem with different objectives, expectations and rights. The tribal cultures

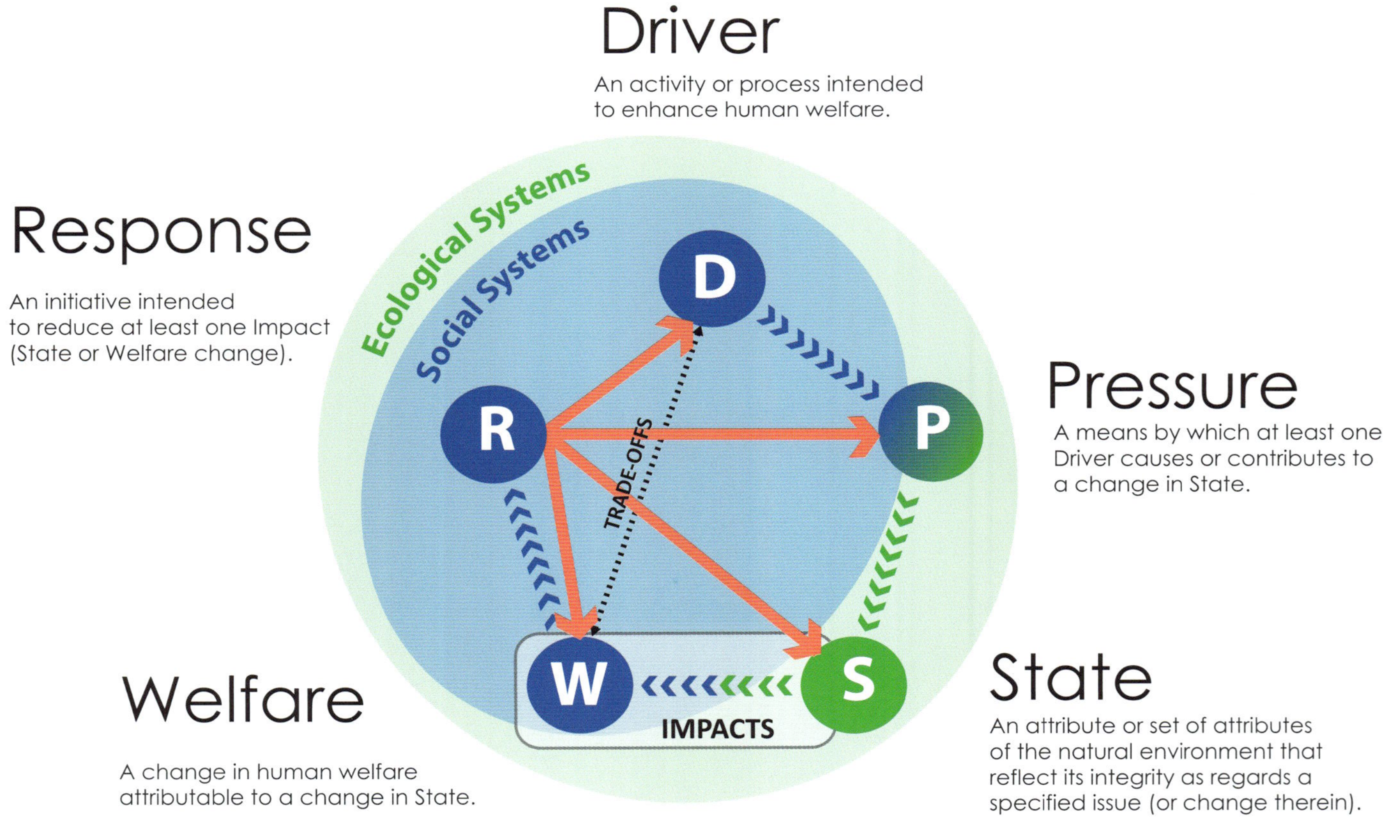

**Figure I.2** The DPSWR framework with definitions for each information category.

## Driver
An activity or process intended to enhance human welfare.

## Pressure
A means by which at least one Driver causes or contributes to a change in State.

## State
An attribute or set of attributes of the natural environment that reflect its integrity as regards a specified issue (or change therein).

## Response
An initiative intended to reduce at least one Impact (State or Welfare change).

## Welfare
A change in human welfare attributable to a change in State.

Ecological Systems

Social Systems

TRADE-OFFS

IMPACTS

P

S

D

W

R

and their dependence on the salmon developed over millennia and the river system was used to sustain their civilization. The non-tribal commercial fishery developed to support larger and more widespread markets. The federal electrification schemes (prompted by an international failure in the economic system) were pivotal to modern economic and social development of the region. Chapter 1 focuses on Drivers of ecosystem change.

Human uses of the environment place **Pressures** on the system which alter the system in different and interacting ways. Pressures are the ways drivers place demands upon an ecosystem. These demands are the interface between the social and ecological components of the system. Fishing has affected stocks directly, both through the traditional dip-netting techniques of the tribes and latterly through more modern techniques. Hydroelectricity has changed the configuration of the river, altered habitat types and made it more difficult for the salmon to run up the river. Chapter 2 examines how Drivers exert different Pressures on the environment and how seemingly similar systems may respond in different ways to similar Pressures.

The pressures caused by human activities result in changes to the ecosystem **State** which is itself constantly changing and adapting. Unpredictable (**stochastic**) processes play a major role in ecosystem functioning. Management of the system must therefore also be adaptive. State changes are the changes in the ecosystem resulting from the Pressures (i.e. ecosystem impacts). Separating the effects of natural fluctuation from those of anthropogenic activities and responding to both natural and man-made changes in the system will require continuing management efforts which must change along with the system. Annual salmon recruitment, even under unharvested conditions, is subject to the fluctuations in oceanic water temperature caused by the PDO. Achievable targets for fish recruitment must therefore take these oscillations into account and adjust accordingly. The ecosystem itself is adapting to its new configuration. For example, the California sea lions have adapted to the presence of the dam and learned the locations where prey may be easily caught. Chapter 3 asks the question how human values relate to identifying a desirable environmental State.

Changes to the ecosystem have had effects on human **Welfare**[2] both in its broad sense of well-being and in the narrower monetary sense. There

---

2  Note that both State and Welfare are different types of what are sometimes called Impacts. The term State refers to environmental impacts while Welfare refers to economic impacts.

are winners: hydro-electricity has brought an unprecedented degree of comfort and development to the region; and losers: communities supported by commercial salmon fisheries have lost their livelihoods. The tribes have suffered a bitter loss of their traditional ways, both through the initial ravages of disease following first contact, and subsequently from man-made changes to the ecosystem in which their culture evolved. Environmental activists are angered by the deaths of marine mammals. Chapter 4 introduces the concept of **ecosystem services** and their classification and reviews some common criticisms of these ideas.

Successful environmental management is dependent on the institutions and organizations that support governance. The **Response** must be focused on the elements of the system that can be effectively managed. The Response to a particular problem may be directed towards any of the other elements (D, P, S or W) in an effort to achieve a balance between the benefits of economic and social development and the ecosystem costs. In the Columbia River example there are highly political aspects to the response concerning the rights of indigenous peoples, the livelihoods of commercial fishermen and the protection of charismatic marine mammals. Management of the system occurs within a legal and institutional framework involving federal, state and tribal governments. Treaties govern the relationship between the tribes and the federal government. State governments work in tandem to restore the salmon, and state laws determine the levels of catch allowed on an annual basis. At the federal level environmental laws for the protection of cherished species regulate the culling of marine mammals. Management efforts are directed at different aspects of the whole system; some regulate human activities, others act directly to restore the environment. Chapter 5 examines how response may be focused on different elements of the system.

Chapter 6 describes an example of a DPSWR case study examining water resources in Al Jabal Al Akhdar, Oman with some comments on the practical challenges of carrying out the analyses, while chapter 7 draws some general conclusions and messages.

## Suggested Reading

Cooper, P. 2013. Socio-ecological accounting: DPSWR, a modified DPSIR framework, and its application to marine ecosystems. *Ecological Economics* 94: 106–115.

Campbell, S. K., and Butler, V. L. 2010. Archaeological evidence for resilience of

Pacific Northwest salmon populations and the socioecological system over the last ~7500 years. *Ecology and Society* 15(1): 17. [online] URL: http://www.ecologyandsociety.org/vol15/iss1/art17/

Lackey, Robert T. 2000. Restoring wild salmon to the Pacific Northwest: chasing an illusion? In: Patricia Koss and Mike Katz (eds). *What We Don't Know about Pacific Northwest Fish Runs – An Inquiry into Decision-Making.* Portland State University, Portland, Oregon, 91–143.

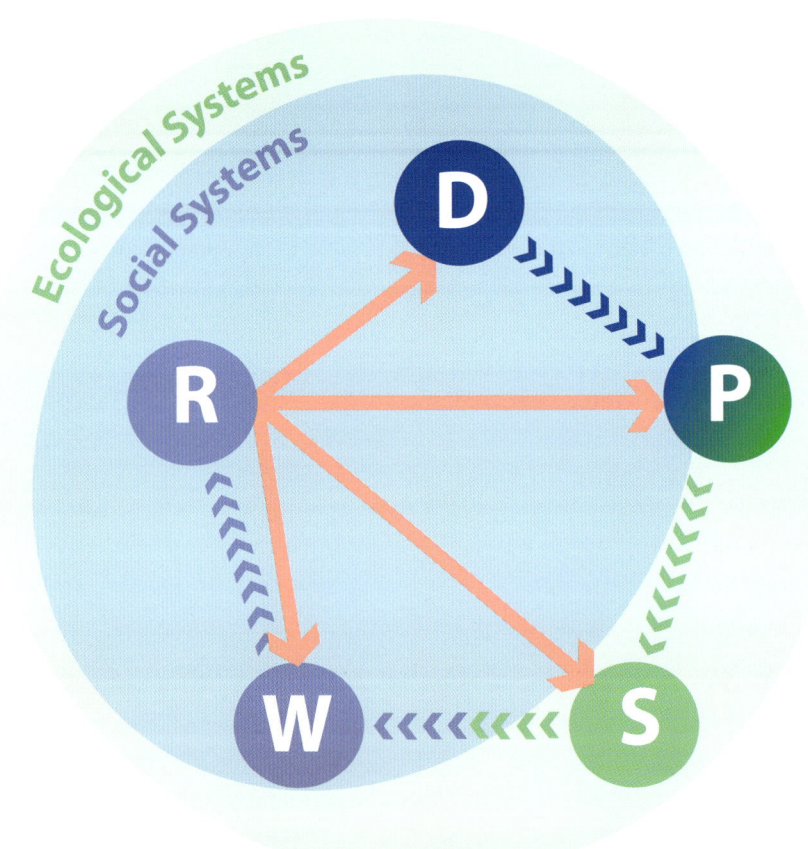

# 1 We have met the enemy and he is us

Meeting the wants and needs of the present without compromising our ability to meet these needs in the future is at the heart of any concept of sustainable development.[1] Drivers are, by definition, the activities or processes that are intended to meet human desires (i.e. produce human welfare). While specific practices, for example fishing, farming and industrial activities (which are directly used to satisfy the needs of society) are the immediate Drivers which place Pressure on the natural environment, these are in turn controlled by more fundamental underlying Drivers such as the urge to reproduce and the urge to live more comfortable lives (i.e. in economic terms to maximize utility). Ultimately the scale of the underlying Drivers dictates the scale and extent of the immediate Drivers and the scales of the Pressures placed on the environment. The science of ecology has long been focused on understanding these **Underlying Drivers** and their consequences for the planet.

On a finite globe beset by human-induced environmental problems, one obvious question is: how many people can the planet support, or what is the human **carrying capacity**[2] of the planet? Estimates of global human carrying capacity have taken several approaches, based on population density data; on mathematical fitting of historical population; and on assumptions of a single limiting factor, amongst others. These estimates vary widely in their results from less than 1 billion to over 1000 billion and have not tended to converge over time (Cohen, 1995); in other words nobody knows. The carrying capacity of the planet depends on the amount of resources needed to support the human population. The level of resource extraction (or the

1  The term sustainability is used to mean so many different things in so many different fields that it has almost lost any real meaning.
2  The concept of carrying capacity has its roots in rangeland ecology and management of the 1890s and has long been accepted in biology (Young, 1998).

level of Pressure on the environment) is dictated by the size of the human population (P = Population), the amount of resources extracted to meet the needs of each individual (A = Affluence) and the efficiency of the technological means (T = Technology) by which these resources are extracted. The relationship between Population, Affluence and Technology to environmental impact is formalized in the IPAT identity,[3] I = PAT (Ehrlich and Holdren, 1971).

Under the DPSWR, Drivers are activities or processes intended to enhance human welfare. Population and Affluence are two of the underlying Drivers of environmental damage. These underlying Drivers result in the demands that drive economies to produce goods and services. In turn, individual economic sectors are the immediate Drivers that place direct Pressures on ecosystems. For example, the demand for fish results in fishing (an economic sector) which places Pressure on the environment through dredging and trawling, which damage seabed habitats as well as altering the structure of aquatic food-webs.

Perhaps the most fundamental of the underlying drivers is the urge to reproduce. Population is a controversial and sensitive topic, both a personal and highly political issue. This urge is engrained by evolution in the human psyche and enshrined as a fundamental human right (United Nations, 1948). Since the Industrial Revolution and its attendant advances in technology and medicine, many of the evolutionary factors limiting growth of human populations have been practically removed for many people. Modern medicine has reduced infant mortality and increased longevity, while the ready supply of electricity enables us to live under extreme climatic conditions, allowing comfortable lives in inhospitable areas[4]. As a consequence, the last two centuries have seen an unprecedented proliferation of human beings around the planet.

Figure 1.1 illustrates a history of global human population from the Industrial Revolution onward along with projected future population. The two most striking features are the continued upward trend, the increasing slope of the curve from the Industrial Revolution to the middle of the twentieth century and the uncertainty of future projections (in blue). The

---

3 The I of IPAT is less rigorously defined than the terms under DPSWR. It may be conceptualized as environmental damage, a mixture between Pressure and State change.
4 Permanent settlement is now even possible at the South Pole.

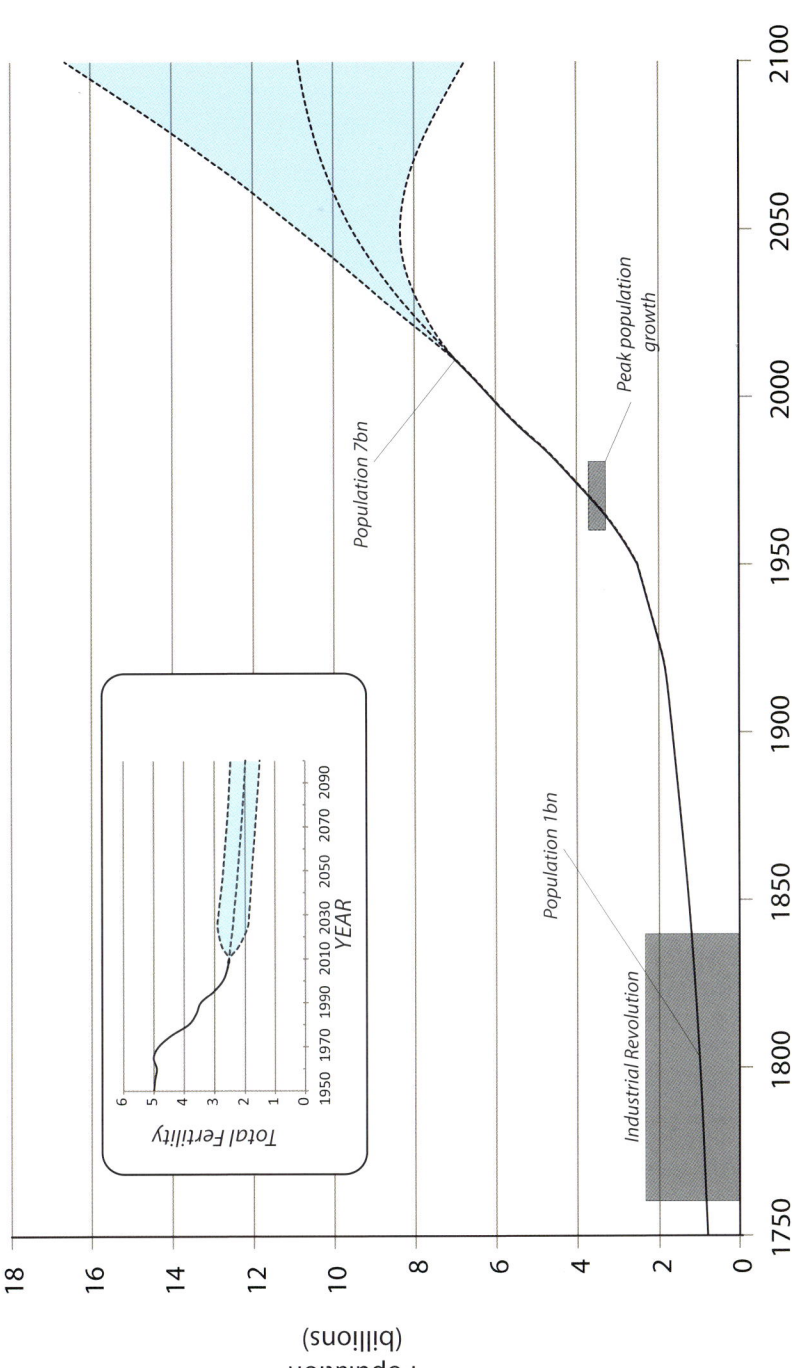

**Figure 1.1**  Time series of human population and fertility rate since the Industrial Revolution indicating major demographic milestones.

recognition of a potential gulf between the needs of the human population and the natural resources required to meet these has been around for centuries. Thomas Malthus (1798) first identified the gulf between the apparent exponential growth in human population and the arithmetic growth in the production of food, 'I say, that the power of population is indefinitely greater than the power in the earth to produce subsistence for man'. Malthus foresaw the potential for disease, famine and war as a consequence of the mismatch. The twentieth century saw the first doubling of global population within a human lifespan. By the late 1960s and early 1970s (and the publication of the IPAT identity), the all-time peak in human population growth was causing increasing alarm. This concern over population was instrumental in the birth of the modern environmental movement. There was increasing popular awareness of the growth of human population and its environmental effects, stimulated in part by publication of the *The Population Bomb* (Ehrlich, 1968). Many of the leading figures in the environmental movement were advocates of direct coercive population control.[5] They espoused solutions such as the conditional provision of aid based on enforced contraception; some explicitly argued for the suspension of aid to developing countries in the face of growing human population, arguing that it was better to leave the poor to die in a resource-constrained world (Hardin,[6] 1974).

The largest-scale, most prominent and perhaps most controversial example of the types of coercive population policies advocated by the early US environmental movement is that of China's family planning policy (popularly known as the one child policy). China's one child policy was responsible for the most rapid change in fertility rates at a national level ever observed. Following the famines in China of the early 1960s, by the late 1970s concerns about the economic consequences of unchecked population growth and the carrying capacity of the nation led to the introduction in 1979 of the one child policy. The Chinese State Family Planning bureau set targets for the maximum fertility rate at an average of 1.2 children born

5 The title of this chapter appeared on a poster for the first Earth Day in 1970, considered to have been the birth of the modern US environmental movement. The phrase is intended here in two senses; firstly, that human population growth causes environmental problems, and secondly that boundaries of ecology and morality are risky political territory.
6 The links between the environmental movement and the eugenics movement are often overlooked. Hardin was director of the American Eugenics Society 1971–1974. Julian Huxley (grandson of 'Darwin's Bulldog', T.H. Huxley) during his tenure as president of the British Eugenics Society was a founding member of the World Wildlife Fund.

per woman. During the early days the policy was enforced through 'shock drives' including mass education programmes, as well as enforced end of year abortion and sterilization campaigns. These included mandatory insertion of intra-uterine contraceptive devices for women who already had a child as well as abortion for unauthorized pregnancies and sterilization for couples with two or more children (Wang, 2012). In 1983 there were almost 88 million sterilizations and over 14 million abortions in China. Since its inception it is claimed that the policy has prevented 400 million births. The policy succeeded in reducing fertility but led to conflicts between the citizens and governments as well as international criticism based on concerns over human rights.

While the policy was remarkably effective in achieving its goal, with total fertility (children per woman) dropping from almost 2.9 to 1.6, it was and is highly controversial for its social implications, not just in terms of the enforced sterilizations and abortions but also in terms of the unintended demographic consequences of the programme. The policy has unintentionally favoured sex selective abortions (though the practice is illegal). Chinese families, for cultural reasons, tend to select for male children. As a result the youngest generation of men and boys (under the age of 20) exceeds the number of females of the same age by about 32 million. The full social consequences of this gender imbalance have yet to be felt and the Chinese government has embarked on a programme to redress the imbalance in future generations (Peng, 2011).

China provides a sobering example of the effects of population control policies, but it is an exception to the general international pattern of demographic changes. **Demographic transition** is a phenomenon involving a shift from high rates of birth and death to low birth and death rates associated with economic development and higher levels of education. When nations reach a certain level of affluence, population growth tends to decline. Figure 1.1(inset) illustrates the declining global fertility rates over the past half century and the predicted future declines based on UN demographic projections. Highest birth and death rates are generally associated with the poorest countries. The current global distribution of wealth amongst the world's 7 billion people (often called the 'champagne glass' distribution) and the average total fertility for each billion are illustrated in Figure 1.2. Increased levels of economic welfare and education generally lead to a

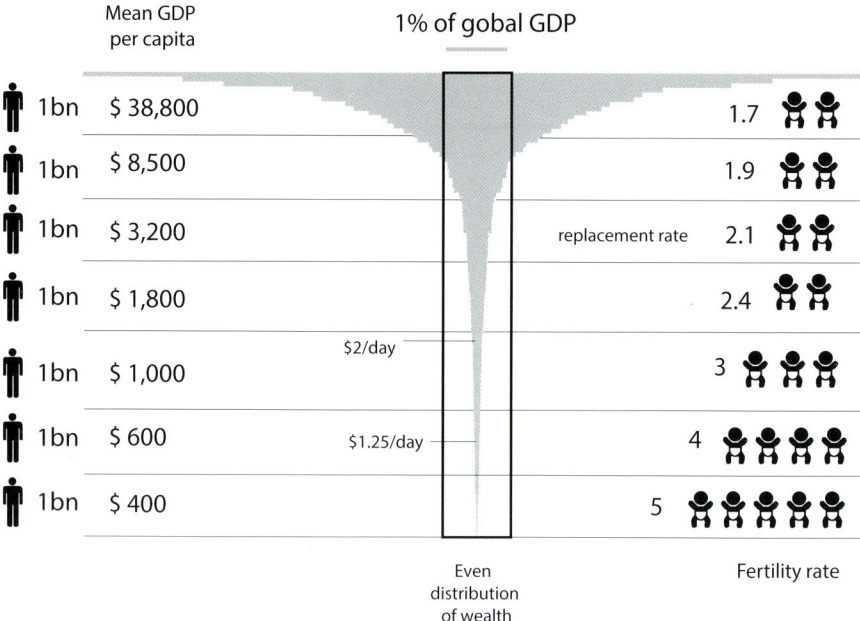

**Figure 1.2** Global distribution of wealth, mean GDP and mean fertility amongst the world's seven billion population.

decrease in both birth and death rates, with the decline in birth rate occurring after the decline in mortality rate, leading to a time lag in the stabilization and reduction in levels of population. At present population growth rates are at or below replacement rates (about 2.1 with current medical technology) in the world's richest three billion but remain above replacement rate in the poorest four billion. The coming decades thus present an enormous challenge; although population growth rate has started to decline, population itself still continues to grow, placing increasing pressures on the planet's biosphere. While the most dramatic population growth rates are perhaps now in the past, the challenge facing modern society is how to achieve the levels of affluence and education to bring about demographic transition globally while meeting the growing needs of the existing, growing population.

Unfortunately population is not the only major Driver of environmental change. The degree of Pressure placed on the environment, and the capacity of the planet to sustain different levels of population depends, of course, on the levels of resource use required to meet individual needs, while these in turn depend on the stewardship and management practices of the economic sectors meeting these needs. The **ecological footprint** is a method developed to measure environmental Pressures, or human demands on the Earth's ecosystem (Wackernagel et al., 2002). The ecological footprint is an indicator comprising environmental impact measures for cropland, grazing, forest fishing, carbon and built up land. It is expressed in global hectares and is a measure of how much of the biologically productive area of the Earth an individual, population or activity requires to produce all the resources it consumes and absorb all the waste it generates using prevailing technology (www.globalfootrprintnetwork.org). The composition of the ecological footprints for several nations is shown in Figure 1.3. Ecological footprints are used to estimate how much of the supply of biologically productive land and sea it would take to support the world's demands on an annual basis. They can be measured at the individual, national or global levels and can be used to compare the levels of environmental resource use between nations. Ecological footprints can be contrasted with **biocapacity** (also measured in

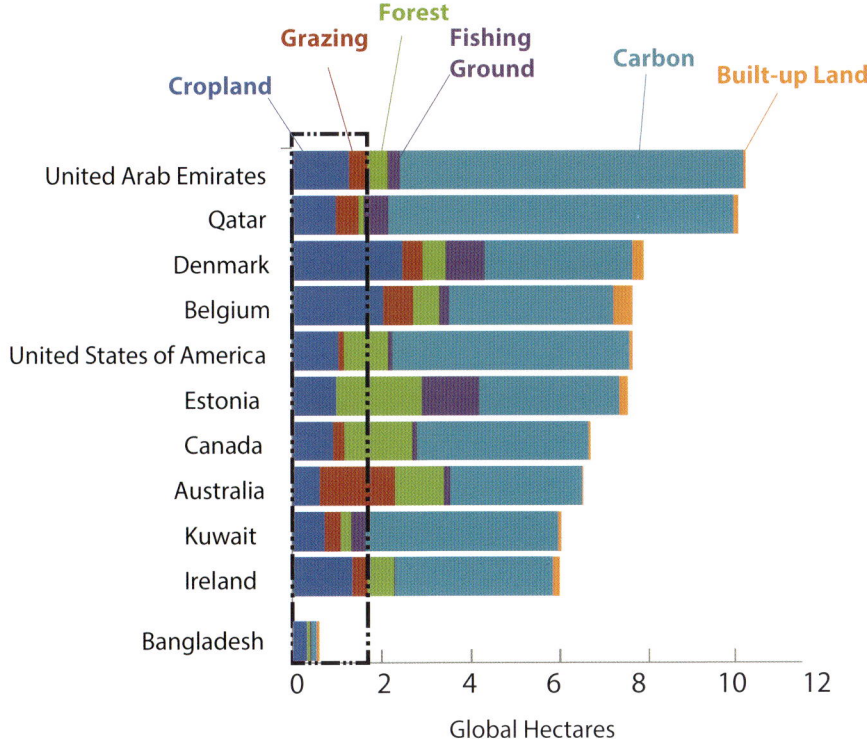

**Figure 1.3** Nations with the highest 10 ecological footprints. Bangladesh is also shown for comparison.

global hectares), a measure of the capacity of ecosystems to produce useful biological materials to meet the needs and wants of society. Details of the method, as well as an international comparison of results, are published in the Ecological Footprint Atlas (Ewing et al., 2010). The most recent synthesis indicates that by 2007 the global human population was annually consuming resources equivalent to 1.5 Earths, that is, that the resources consumed in 2007 would take 1.5 years to be replenished. The current ability of society to exceed the biocapacity of the planet is due to our reliance on existing stocks of natural resources (e.g. freshwater reserves, fossil fuels) which have built up over geological time. The importance of fossil fuels (represented by the carbon footprint) in contributing to the overall footprint of wealthy nations is clearly illustrated in Figure 1.3.

Figure 1.4 illustrates a relationship between economic development (as GDP per capita) and ecological footprint. Biocapacity per capita means the number of global hectares available to each person at current population levels. These data show that, in general, the wealthier the nation the greater the margin by which biocapacity is exceeded, and that only poorer nations are living within their ecological means (at or below biocapacity). All European nations, as well as the US and Canada, are exceeding biocapacity while the majority of African and many Asian countries are at or below capacity.

Taken in combination, the Drivers of population and affluence perhaps represent the two most significant challenges to the future of global ecological integrity. Reducing poverty, increasing affluence, welfare and education offers global society the chance to achieve demographic transition as well as improving the living standards of the poorest people (pushing countries further to the right on the X axis of Figure 1.4). At the same time, global footprints illustrate that developed nations are living beyond the natural limits dictated by the ecological functioning of the planet, and that to live within the long-term boundaries of sustainability set by the productivity of the planet **economic transition** is required (a reduction in ecological footprint, down the Y axis of Figure 1.4). These combined phenomena offer an uncomfortable dilemma. In the absence of new technologies to radically increase resource efficiency, control of population or reductions in the level of affluence are inevitable. The unpalatable history of the one child policy provides an example of the former, while voluntary reductions in the latter

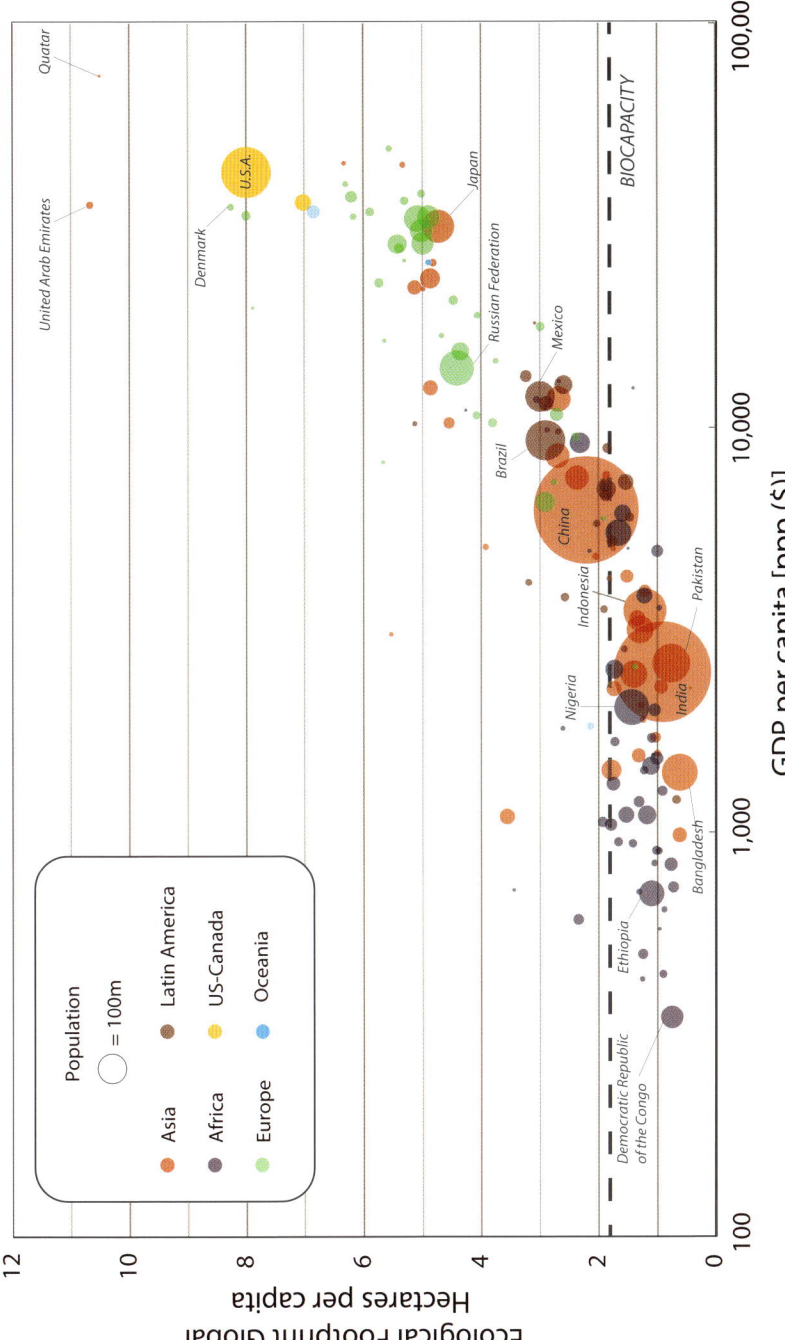

**Figure 1.4** Ecological footprint plotted against GDP. Bubble size indicates population. Continents are colour coded; the dashed line shows biocapacity.

are untested. At present growth is the foundation of modern economies, and concepts such as steady state economies (Daly, 1973) and de-growth (Assadourian, 2012) are antithetic to any established modern economic practices, though they have some currency in academia. This dilemma is not new – it has been recognized and fiercely debated since the first conceptualization of IPAT and before. The original authors of IPAT (Ehrlich and Holdren, 1971) stressed the importance of population while others decried the wastefulness of contemporary society (Commoner, 1972) but foresaw technological solutions.

Technology in IPAT was initially thought of as contributing to overall Pressures, associated with the effluents of industrial and technological development. Technology is now also identified as a factor that can reduce the Pressures associated with specific Drivers (*see* Waggoner et al., 2002 for a formal contemporary treatment of technology in the context of IPAT identity). The famines envisaged in *The Population Bomb* did not occur. This was due (at least in part) to intensive programmes of agricultural research, the 'Green Revolution', which achieved a tripling of cereal crop production (concurrent with a doubling of population) over the past 50 years with just a 30% increase in area of land cultivated (an economic transition). This technological advance saved land from conversion to agriculture, reduced malnutrition levels and improved health for millions (Evenson and Gollin, 2003; Pingali, 2012).

While the underlying Drivers of economy and population are the root cause of many environmental problems, their fundamental nature and their size make them difficult to address on environmental grounds. Direct responses such as reducing population or curbing economic growth have profound implications in terms of human welfare, which make them not only politically unpopular but also socially undesirable. While economy and population are underlying Drivers (with highly political and philosophical issues surrounding any potential management actions), responses directed at more immediate Drivers can represent a more direct way of solving a particular environmental problem.

The **immediate drivers** of environmental problems are the ways in which the wants and needs of society are met through particular activities or economic sectors. For example, the need for food may be met by a number of different sectors; hunting, fishing and foraging meet demand for food through harvesting of wild natural resources. Agriculture and aquaculture meet the same demand through exploitation of cultivated stocks of the same resources, which may be more readily controlled and optimized. In practical terms for a specific environmental problem, solutions may be more readily identifiable by addressing particular economic activities or immediate drivers. Improving food production through the technology of the Green Revolution posed less of a moral and ethical dilemma than making decisions on the number of people allowed to eat. The rest of this book deals with the immediate Drivers and the Pressures they place individually on different ecosystem components. Nevertheless, the profound moral questions touched on in this chapter provide the general context about the depth and importance of the underlying problems of human-induced environmental change and the uncomfortable truths of environmental advocacy.

Ultimately the levels and types of human activity have profound consequences for the health of the ecosystems that provide the resources. While demographic transition and the Green Revolution may have meant that the direst predictions associated with population have not (yet?) occurred, global population continues to grow and natural resources continue to be depleted ever more rapidly. The underlying Drivers of environmental problems clearly pose enormous moral and political challenges internationally. The future size of human populations and levels of affluence are uncertain. Whatever actions are taken will require moral judgement decisions, resulting in winners and losers. Ecological concepts such as carrying capacity are of limited use when considering complex and adaptive social-ecological systems with the ingenuity to change behaviours under altering resource constraints. While the science of ecology can alert us to the natural resource consequences of expanding human populations and increasing affluence, it does not provide a supporting ethical framework on which to base management decisions.

Managing underlying drivers, controlling population, or limiting levels of affluence are intensely difficult and highly political questions. While there must be limits to the carrying capacity of the planet, these are not fully known and are dependent on human patterns of population and consumption. From a pragmatic environmental management perspective, immediate Drivers, the individual economic sectors that place direct pressure on the environment may be an easier target for management effort than the underlying Drivers with their associated economic, political and ethical implications. The following chapter illustrates an immediate Driver, agriculture, the Pressure it produces and the resulting changes in environmental State.

## Suggested Reading

Cohen, J.E. 1995. Population growth and Earth's human carrying capacity. *Science* 69: 341–346.

Ehrlich, P.R. and Holdren, J.P. 1971. Impact of Population Growth. *Science* 171: 1212–1217.

Hardin, G. 1974. Living on a lifeboat. *Bioscience* 24: 561–568.

Peng, X. 2011. China's demographic history and future challenges. *Science* 333: 581–587.

United Nations. 1948. Universal Declaration of Human Rights (adopted 10 December 1948 UNGA Res 217 A(III) (UDHR).

Wackernagel, M., Schulz, N.B., Deumling, D., Linares, A.C., Jenkins, M., Kapos, V., Monfreda, C., Loh, J., Myers, N., Norgaard, R., and Randers, J. 2002. Tracking the ecological overshoot of the human economy. *Proceedings of the National Academy of Science* 99: 9266–9271.

Wang, C. 2012. History of the Chinese Family Planning program: 1970–2010. *Contraception* 85: 563–596.

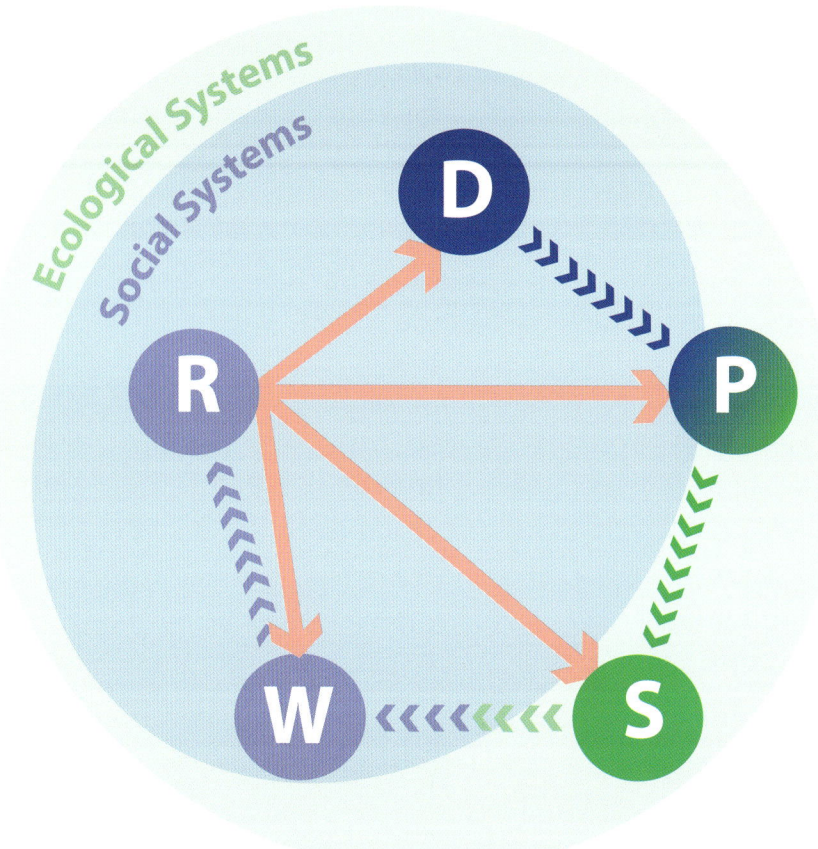

# 2 Feeding and excretion: eutrophication in Eastern Europe

As we have seen in the previous chapter, the Malthusian predictions of the 1960s and 1970s have not so far been borne out, and mega famines have been averted, in part due to the Green Revolution, in part due to demographic transition and the resulting slowdown in population growth. Nevertheless, while the threat of catastrophe still looms large, the efforts to meet the growing needs and wants of an expanding population still pose environmental challenges. Food security (another fundamental human right), a need as distinct from a want, can be seen as a minimum level of affluence for any reasonably stable social system.

Agriculture is the main economic sector, the immediate Driver, responsible for putting food into the mouths of the billions. Global agriculture may be conceptualized as a system with the aim of converting nutrients essential to the functioning of human bodies into digestible forms for transfer from the environment into those bodies. As described above in the previous chapter, there are an increasing number of humans to feed, and this necessitates the greater production of food and greater inputs of these nutrients. Having already grown three-fold in the last century it is estimated that a further 70% growth in agricultural production will be required to meet demand for food by 2050 (United Nations, 2013). Figure 2.1 illustrates the global coverage of agriculture (c.2000) showing that deserts, forests, mountain ranges and areas of ice and tundra are the only major parts of the land not put to agricultural use. The rapid increase in food production over the past 40 years was met principally by intensification of agricultural techniques, not by expansion of agricultural lands. Meeting the predicted increase in demand will require further intensification. Over the last 60 years agricultural intensification has been achieved partly through the development of new strains of grain, but more particularly through the increased utilization of fertilizers

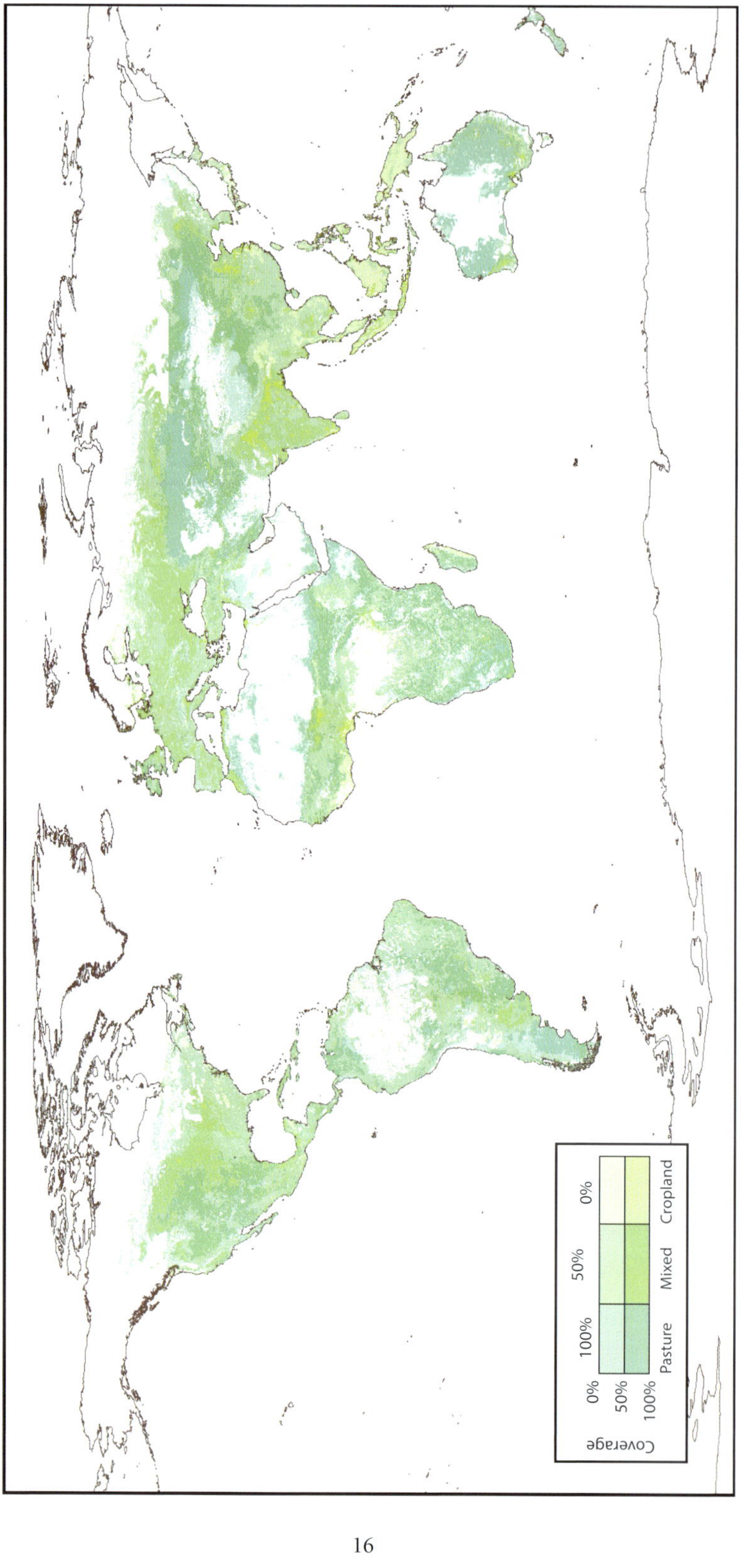

**Figure 2.1** Global distribution of agriculture: data from Ramunkutty et al. (2010).

containing nitrogen and phosphorus. The scale of agricultural production and the vast quantities of nutrients required to meet this production have led to the domination of global biogeochemical cycles of nitrogen (Vitousek et al., 1997) and are considered by some to have exceeded safe limits (Röckstrom et al., 2010).

Modern Europe provides an excellent example of the challenges of maintaining food security through intensification of agriculture. Following the horrors of the Second World War,[1] Europe licked its wounds and resolved never again to fall into such conflict. It turned instead to a 'labour of peace', the European project. It was initially designed to make another war between France and Germany an impossibility. A common market was established in 1957 under the Treaty of Rome. The initial vision for Europe focused on heavy industry, but during and after the war Europe had experienced severe food shortages. By 1962 the Common Agricultural Policy (CAP) had been put into practice with the aims of achieving food security and preventing hunger in the region through modernization of farming techniques and ensuring good prices for farmers. The policy has achieved and exceeded the production levels for food security in Europe ever since. At the same time, on the other side of the Iron Curtain the USSR, having also suffered terrible hunger before and after the Second World War, continued its programmes of agricultural intensification. In Eastern Europe and western Russia, agriculture has over the last 50 years been a major Driver and has placed significant Pressures on the environment, causing unintentional and undesirable changes in environmental State.

Eastern Europe and western Russia straddle two major drainage basins (Figure 2.2). The Baltic Sea receives freshwater inputs from the nations of the north including Poland, Lithuania, Latvia and Estonia (former Soviet nations) as well as much of Sweden and Finland and parts of Denmark and Germany. To the south lies the Black Sea, which receives waters from the Danube (the most international river basin in the world) from the northwest as well as Turkey, Georgia and the Russian Federation. Combined, the Baltic and Black Sea basins drain an area from the Arctic to Asia. The region is also

---

1 Perceived over-population was the underlying Driver of 'Lebenshraum' (or living-space), the German policy of expansion, which led to World War II. *In the briefest outline, Germany's economic position is as follows: 1. We are overpopulated and cannot feed ourselves from our own resources…* ' (Hitler, 1936).

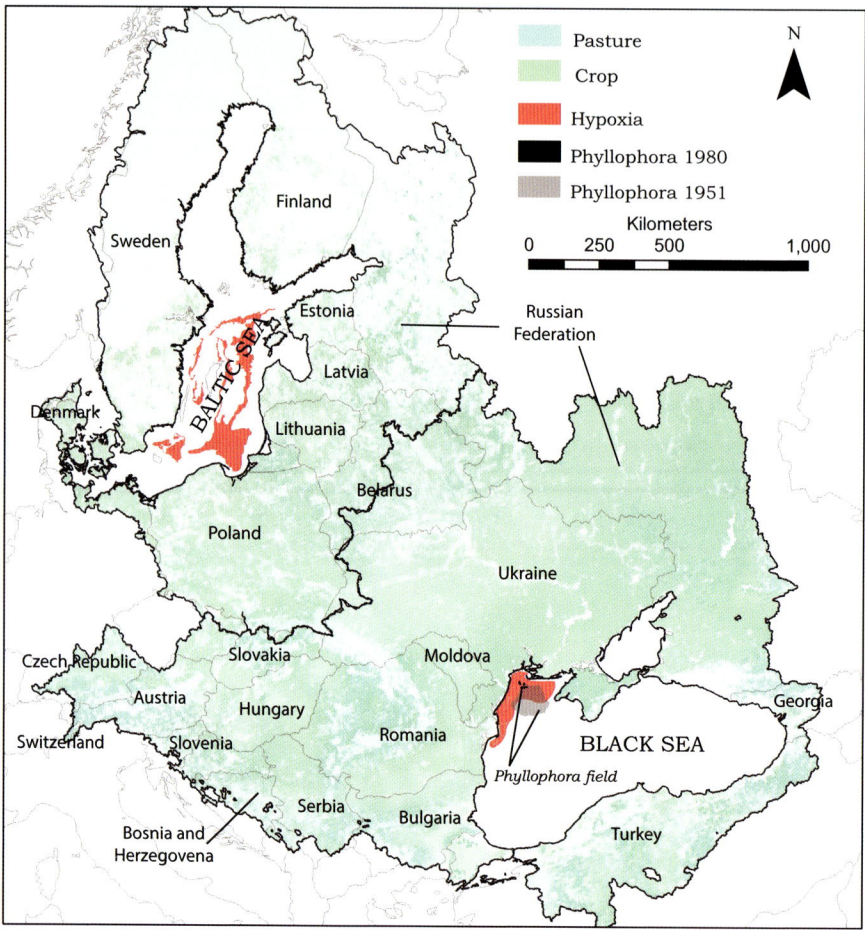

**Figure 2.2** Coverage of cropland in the Baltic and Black Sea catchments: data from Ramunkutty et al. (2010). Red areas indicate anthropogenic hypoxia.

characterized by great diversity in language, culture and economic prosperity. In the past half century the region has seen many re-arrangements of political boundaries, from the collapse of the Soviet Union to the break-up of former Yugoslavia, and most recently the re-drawing of the Ukrainian border.

Despite the diverse culture and the shifting international political currents, the overall pattern of environmental Pressure has been similar throughout the area. Figure 2.3 shows cereal production and fertilizer consumption from the period 1961 to 2002 in the two catchments.

Cereal production grew relatively steadily from the 1960s onward. Over the period shown in the graph a total of 6.9 billion tonnes (~5km$^3$) of cereals

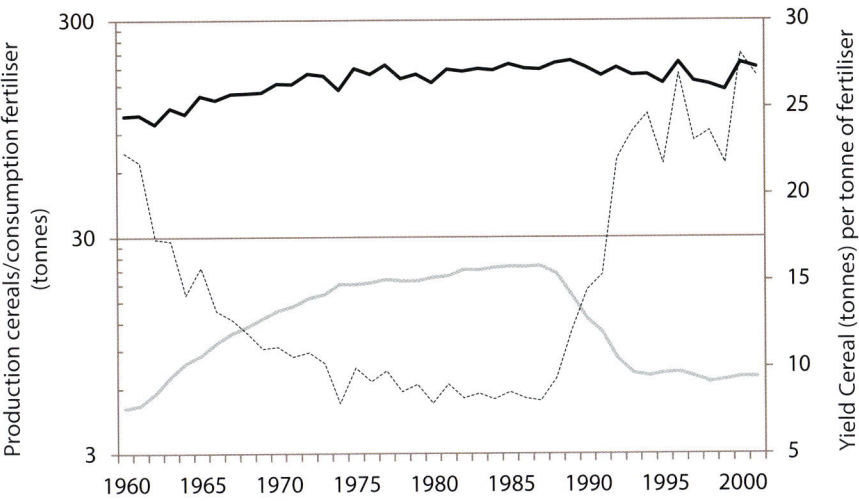

**Figure 2.3** Time series of fertilizer distribution (solid grey), crop production (black), and yield (dashed grey) for the Black and Baltic Sea catchments 1961–2002.

were produced in the region and the annual production increased by about 50% (or 55 million tonnes per year) in that time. Over the same period the consumption of fertilizer in the regions quadrupled. From 1962 until the mid-1980s there was a strong relationship between increasing fertilizer consumption and cereal production. From the start of the period until the 1980s the yields of cereal per tonne of fertilizer steadily decreased as fertilizer consumption continued to increase. Diminishing returns on the efficiency (ratio of crop production to fertilizer consumption) led to a gradual decrease in the rates of fertilizer application in the late 1980s. The steep decline (note the log scale of the graph) from 1989 until the mid-1990s was the result of the collapse of the centralized economic system of the USSR and the other nations of the Eastern bloc. The increase in ratio of application to production following the collapse of the USSR indicates the overuse of fertilizer in the earlier period. Since the mid-1990s the yields of cereal per tonne of fertilizer consumed have quadrupled. The continued high levels of cereal production (relative to the earlier period) also suggest that the cereal production in the region is still benefiting from the historical application of fertilizers. While modern intensive farming practices have successfully provided food for over half a century, the waste of fertilizer has had dramatic unforeseen consequences for the two semi-enclosed seas of Eastern Europe.

Eutrophication is the process of enrichment of water by nutrients, especially compounds of nitrogen and/or phosphorus, causing accelerated growth of algae and higher forms of plant life to cause undesirable disturbance to the balance of organisms present in the water and the quality of the water concerned (EU, 2000).

Each part of the agricultural conveyor belt from farmland to fork involves nutrient waste that can cause eutrophication. While much of the nutrient applied as fertilizer enters crops (and from there the human body) a proportion remains on the land. The fertilizer not incorporated into the crops eventually runs off into streams and rivers, into lakes and groundwater, and ultimately these nutrients arrive at coastal seas. Wastewater discharges from domestic and industrial activities are the second major source of nutrients to aquatic environments. Nutrients in human waste water may also have their origins in fertilizer, but in this case the fertilizer has been taken up by crops and perhaps in turn ingested by animals to produce the meat for human consumption, which ultimately enters the hydrological system.

Figure 2.4 shows a conceptual diagram of eutrophication indicating the major point and diffuse sources of nutrients. Fluxes of nutrients to the aquatic environment are the Pressures[2] placed by agriculture on aquatic environments, and these Pressures cause changes in the environmental State of the water bodies. Excess nutrient loading leads to increased growth of opportunistic algae, including seaweeds (in particular green seaweeds) as well as microscopic phytoplankton. Elevated phytoplankton concentrations reduce the clarity of the water and the penetration of light through the water. The decrease in light penetration reduces light availability to other longer-lived aquatic plants including seaweed species and sea-grasses and may cause them to die. In turn, decomposition of algal material depletes the level of oxygen in the water (resulting in hypoxia) and can cause fish and bottom-dwelling organisms to die.

---

2 Pressures are perhaps the most difficult of the DPSWR elements to conceptualize; they are NOT human activities and they are NOT changes in the environmental State. Pressures lie at the boundary between the social and ecological systems; they are the means by which the activity causes the change in environmental State. Consider the example of a forester felling trees. The Pressure is the axe (or chainsaw). The change in the forest resulting from the action of the axe (Driven by the need for wood) depends on the sharpness of the axe, the force at which it is driven home and the number of times it is used, while the State change depends on the hardness of the wood to which it is being applied.

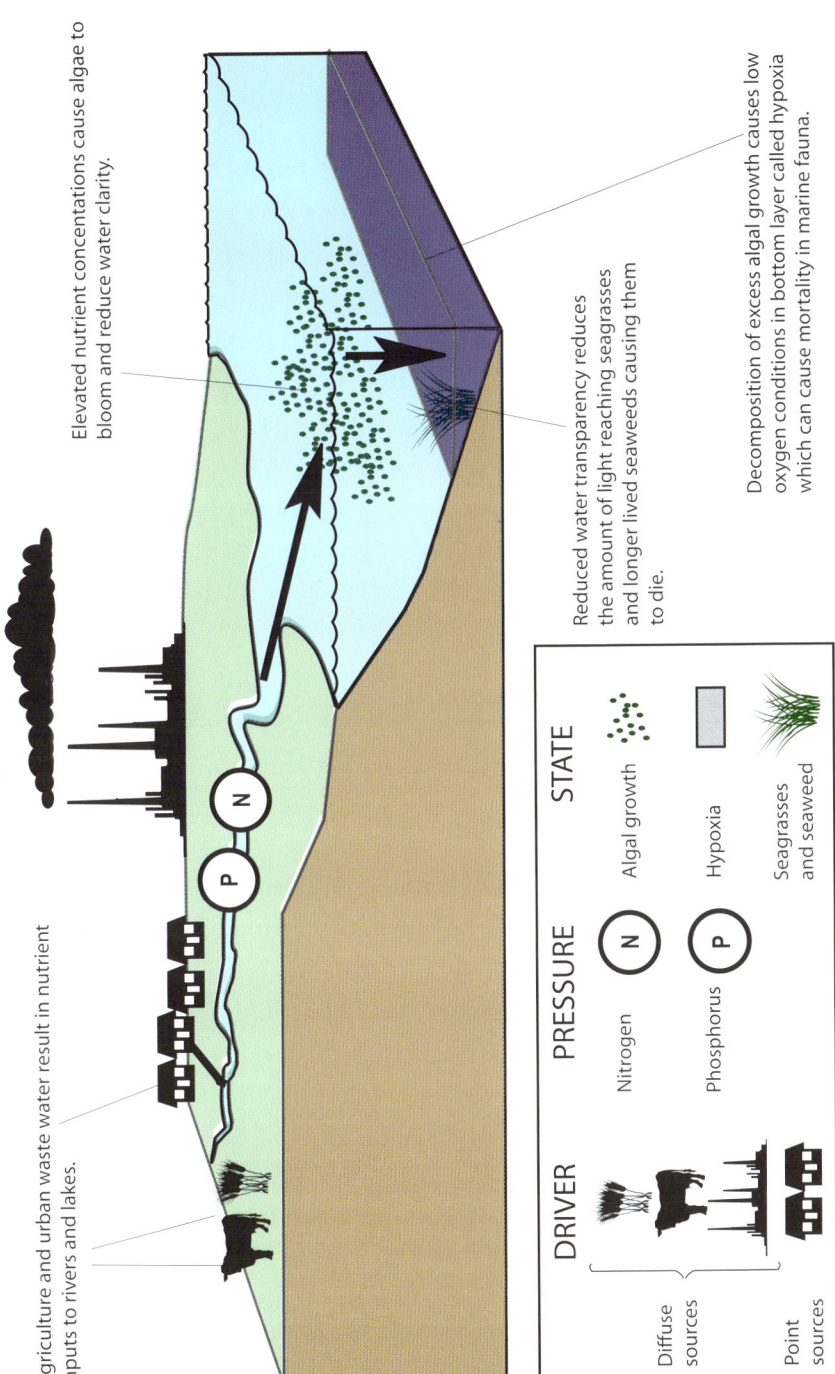

Atmospheric emissions of nitrogen from industry

Agriculture and urban waste water result in nutrient inputs to rivers and lakes.

Elevated nutrient concentrations cause algae to bloom and reduce water clarity.

Reduced water transparency reduces the amount of light reaching seagrasses and longer lived seaweeds causing them to die.

Decomposition of excess algal growth causes low oxygen conditions in bottom layer called hypoxia which can cause mortality in marine fauna.

DRIVER

Diffuse sources

Point sources

PRESSURE

N   Nitrogen

P   Phosphorus

STATE

Algal growth

Hypoxia

Seagrasses and seaweed

**Figure 2.4** Schematic diagram showing the major causes of eutrophication.

The problem of eutrophication in freshwater and coastal environments has been around throughout history in Europe[3],[4]. In modern times the environmental impacts (State changes) associated with agriculture have been observed in each of Europe's regional seas and have often had dramatic consequences. These include the death by hydrogen sulphide asphyxiation of domestic animals, including horses, on the Atlantic coast of France (Menesguen et al., 2010) as well as international legal battles resulting from rows over effluent emissions in the North Sea (European Union, 2012). The capacity of particular ecosystems to absorb the Pressures resulting from human activities depends on the peculiarities of each ecosystem. The expression of coastal eutrophication, the State changes caused by nutrient enrichment, vary depending on the ecosystem characteristics of a particular location (Cloern, 2001).

Having experienced the same historical pattern of agricultural Drivers and Pressures and the abrupt decline in fertilizer application, the Black and Baltic Seas provide an example of eutrophication at large spatial scales and illustrate how ecosystems can respond differently to similar Pressures. Figure 2.5 summarizes some of the main physical characteristics and major Pressures associated with eutrophication of the Black and Baltic Seas.

## The Black Sea

The Black Sea is an almost fully enclosed sea with a connection to the Mediterranean via the narrow (<1km) Bosphorus. Most of the Black Sea has a relatively thin strip of shallow shelf surrounding a deep offshore abyss. The deep abyss is made up of permanently stratified waters which are naturally without oxygen below a depth of about 200m. Eutrophication in the Black Sea has been most dramatic on the widest, north-western part of the shelf at the Danube Delta (straddling the border of Romania and the Ukraine) where the fresh waters of the Danube River meet the brackish waters of the Black Sea. The Danube accounts for 70% of nutrient loads to the Black Sea.

---

3 The biblical account of the first plague visited on the Egyptians describes fish kills, foul smells and water discoloration, all symptoms of eutrophication. '...all the water that was in the Nile was turned to blood. The fish that were in the Nile died, and the Nile became foul, so that the Egyptians could not drink water from the Nile' (Exodus: 7, 20–21).

4 Perhaps the earliest scientific record in the European marine context is that of Adeney, 1908, who recognized the profusion of 'sewage algae' in Dublin Bay.

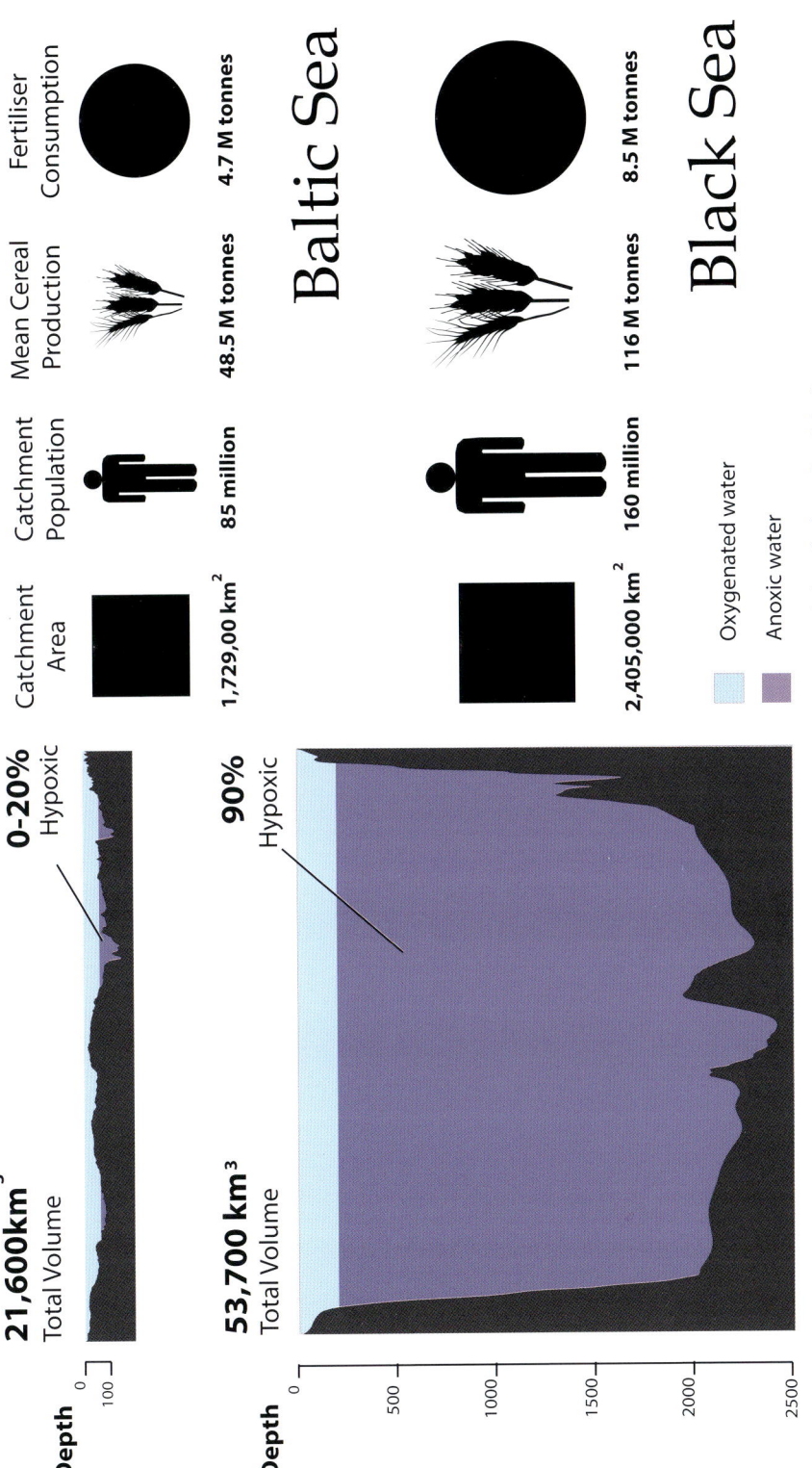

**Figure 2.5** Some social and ecological characteristics of the Black and Baltic Seas.

Four distinct phases of eutrophication in the Black Sea can be identified: the pre-eutrophic phase up to 1970; the early eutrophic phase in the 1970s; the intense eutrophic phase from 1980 to 1995 and an on-going, post-eutrophic phase (Oguz and Velikova, 2010).

In the pre-eutrophic phase the north-western shelf of the Black Sea housed a unique fauna (adapted over 10,000 years to the warm, brackish conditions of the northern shelf). Figure 2.6 shows the distribution of habitat types of the Black Sea's northwest shelf including the unique Zernov's Phyllophora field. This Phyllophora field was a habitat comprising several species of the red algae Phyllophora and occupied an area of 11,000km² on the Ukrainian shelf. The habitat was host to a specially adapted fauna including 118 species of invertebrates and 47 species of fish and also supported commercial harvesting of alginates. In the 1970s (the early eutrophic phase), concentrations of phytoplankton (microscopic marine plants) in the surface water began to increase and water transparency began to decrease. This phase also corresponded with diminished numbers of top predators (through overfishing). Both the pre-eutrophication and early eutrophication were associated with relatively warm and mild climatic conditions, and these conditions are thought to follow multi-annual cycles (Oguz et al., 2006; Oguz and Velikova, 2010).

In the intense eutrophication phase (1980s and 1990s), which corresponded with the most profligate use of fertilizers (*see* Fig. 2.3) phytoplankton concentrations continued to increase, and hypoxic and anoxic events in the bottom waters occurred, contributing to the loss of the complex benthic ecosystems that had previously been present. Reduced water clarity caused by the Pressure of nutrient loading resulted in severe reduction in the extent of the Phyllophora. Mass mortalities of benthic invertebrates were a regular occurrence and the Black Sea was perceived to be in crisis (Mee, 2006). By the early 1990s the field covered only 500km² and the uniquely associated fauna had collapsed.

In the post-eutrophic phase there has been a gradual return to pre-1970s phytoplankton concentrations and the number of hypoxic events and the extent of the hypoxia have diminished. Water clarity and phytoplankton abundance have responded relatively rapidly to the relaxed nutrient Pressures. The Phyllophora field has not shown the same recovery. Despite the reduction in nutrient Pressure the system as a whole has not reverted to

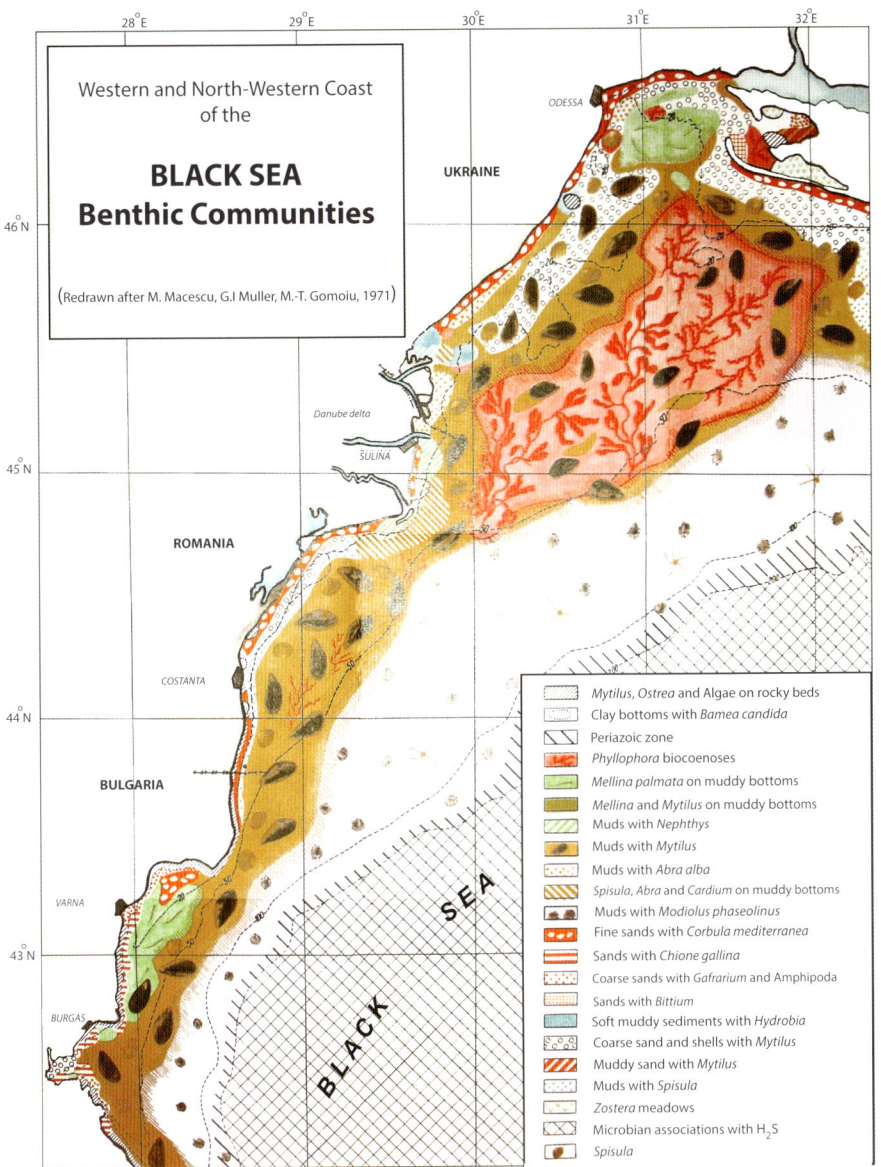

**Figure 2.6** Major habitat types of the western and northwestern Black Sea Shelf. © Marion Gomoiu.

its pre-eutrophic condition. The food web remains considerably altered and appears to have reached an alternative stable State with a dominance of jelly species and regular occurrence of blooms of the phosphorescent alga *Noctiluca scintillans* (Oguz and Velikova, 2010).

## Eutrophication and the Eastern Baltic Cod

The Baltic Sea is relatively shallow, with an average depth of just 55m. Like the Black Sea, the Baltic is almost fully enclosed. It connects to the North Sea via the shallow (18m) and narrow (116km) Danish Straits. This narrow channel limits the exchange of waters between the two water bodies. To the east of the Danish Straits the Baltic Sea is stratified. That is, less dense fresh water lies on top of denser salty water. This layered structure prevents oxygen from the atmosphere from reaching the deeper (more saline) waters. As a result, the decomposition of organic matter in these deep bottom waters causes depletion in the levels of oxygen there. Oxygen is only naturally replaced when dense, oxygen-rich, salty waters are forced into the Baltic from the North Sea and flow over the narrow sill of the Danish Straits to mix with the deep Baltic waters. The climatological and hydrographic conditions required for this exchange of water occur rarely. Historically these inflows occurred approximately every five years, but in the past two decades these inflows of North Sea waters have been sporadic and infrequent, with major events occurring in 1993, 1997, 2003 and again a decade later in the winter of 2011–2012 (Nausch, 2013). In the interim periods between inflow events the Baltic Sea has a tendency to stagnate. Even in the absence of human Pressures, the decay of organic matter in the Baltic gradually reduces oxygen concentrations in the deeper water and causes anoxia in the deeper basins of the Sea. The unique hydrographic characteristics of the Baltic Sea and its long **residence time** make it particularly susceptible to eutrophication.

From prehistory cod has been a mainstay for the peoples living on the shores of the Baltic Sea. Cod are the most valuable of fish stocks in the Baltic and played a pivotal role in the culture and economic life of the region through the Viking and medieval periods on into the modern era (Enghoff, 1999; Barrett et al., 2008). Eastern Baltic cod form a distinct population separated from the western Baltic cod stock and uniquely adapted to the salinity of the area. The density structure of the water is critical to the growth and survival of the Eastern Baltic cod. The cod lay their eggs in the open water. These eggs are heavier than fresh water and sink through the water column until they reach a saltier layer that matches their density. The eggs float passively in these deep, saltier layers until hatching. The success of the spawning in any particular year depends on the amount of oxygen in the dense layers where the cod eggs settle. The amount of this water is called the spawning volume (Pilkshs et al., 1993) and this volume is dictated by the sporadic North Sea inflows.

As in the Black Sea, modern fishing and farming practices (incentivized by Europe's Common Agricultural Policy and Common Fisheries Policy) have resulted in Pressures which have altered the State of the ecological system. The State of the system is crucial to the survival of the cod (and the fishery which it supports).

Eutrophication in the Baltic manifested itself as early as the 1950s in the form of phytoplankton blooms around major urban areas (Finni et al., 2001). The excess algal matter produced by the addition of nutrients to the Baltic Sea ecosystem results in an increased demand for oxygen in the bottom waters, depleting the overall level of oxygen. The extent of hypoxic areas has shown a gradual increase in the Baltic Sea and this has resulted in reduced spawning volumes and reduced recruitment of cod.

There have been substantial efforts to reduce the loads of nutrients to the Baltic. In particular, nutrient loads from point sources (wastewater) have decreased substantially. Diffuse sources, such as agricultural loads, still remain relatively high. Despite the decreases in nutrient loads to the Baltic the symptoms of eutrophication have persisted. Phosphorus has accumulated in bottom sediment and continues to fuel phytoplankton blooms. The ecosystem itself has adapted to the change in nutrient loading. Nitrogen-fixing phytoplankton have come to the fore, fixing nitrogen to exploit the accumulated pool of phosphorus (Vahtera et al., 2007).

The application of fertilizer caused similar Pressures in both systems but the symptoms of eutrophication have expressed themselves differently and present uncertain pathways to recovery. The State changes and subsequent recovery rates of these two seas illustrate a number of important characteristics of ecological systems and their ability to withstand Pressures resulting from the agricultural Driver.

In both systems the current state of the environment is not simply the result of the current Drivers and Pressures but is also affected by legacy effects of past and present Drivers and Pressures (O'Higgins et al., 2014). These include **Memory effects** and **Future effects**. Memory effects occur where past Pressures have caused changes in the current State of the ecosystem (for example, the Baltic still remains hypoxic despite the reduction in nutrients). Similarly, future effects can also be observed in both seas. Here current activities, for example ongoing agriculture, cause Pressure in another ecosystem component (the terrestrial landscape); that is, nutrients currently being applied in the catchment will cause Pressure in these eco-systems in the future. In the Baltic, estimates suggest a lag of about 30 years between application of fertilizer and its arrival in the marine environment (Neumann, 2007). These memory and future effects (or legacy effects) con-strain current and future environmental state and limit human capacity to manage the systems.

In these examples, the expression of memory and future effects is medi-ated by the ecosystem characteristics of each sea. The Black Sea's phytoplank-ton recovered relatively quickly (in less than a decade) after the collapse of the USSR and relaxation of nutrient pressures afterwards. This was caused by the (relatively) rapid exchange of water off the shallow shelf and into the large, deep anoxic layer that acts as a semi-permanent sink for excess nutrients. In the Baltic, despite the huge effort to reduce nutrient loads, the sea's shallow profile, slow flushing times and the adaptation of the ecological components have all combined to produce a longer-lasting memory effect.

Aspects of both ecosystems appear to have reached alternative stable states, demonstrating *resilience* to change. In the Baltic the nitrogen-fixing bacteria now exploit the high phosphorus, low nitrogen conditions so that further reduction in nitrogen has a limited capacity to return the ecosystem to its previous state. In a similar way, despite the return of the chlorophyll to pre-eutrophic levels, the Black Sea food web has also been resilient as a whole and has not returned to its pre-eutrophic state.

The cases of the Black and Baltic Seas illustrate not only how the nutri-ent Pressures caused by human population and agriculture have unintended negative consequences on the environment, but they also demonstrate that these ecosystems respond in complex and unpredictable ways. Apparently random climatological and hydrographic effects determine how the systems recover from anthropogenic shock. In the Baltic this is controlled by the inflows from the North Sea; in the Black Sea it appears to be related to the strength of the North Atlantic Oscillation (Oguz et al., 2006). The State of both ecosystems is linked to atmospheric processes that are unpredictable and not fully understood. Even if the relations between climate, oceano-graphic conditions and biological production are relatively well understood (for example, in the case of the Baltic cod), the climatic conditions and the productivity of the cod lie beyond our control and any management must work under these constraints.

The examples in this chapter have focused on agricultural Pressures and their resulting State changes in two marine environments. Just as the Baltic and Black Seas had different reactions to the Pressure, so any ecosystem with its own unique set of physical, chemical and biological and climatological characteristics will exhibit different ecological State changes and different rates of recovery following relaxation of Pressure. Determining appropriate environmental management strategies must account for these complex and unpredictable phenomena.

## Suggested reading

Cloern, J.E. 2001. Our evolving conceptual model of the coastal eutrophication problem. *Marine Ecology Progress Series* 210: 223–253.

Mee, L.D. 2006. Reviving Dead Zones. *Scientific American* 295: 79–85.

Oguz, T., and Velikova, V. 2010. Abrupt transition of the northwestern Black Sea shelf ecosystem from a eutrophic to an alternative pristine state. *Marine Ecology Progress Series* 405: 231–242.

O'Higgins, T.G., Cooper, P. Roth, E. Newton, A. Farmer, A., Goulding, I., and Tett, P. 2014. Temporal constraints on ecosystem management: Definitions and examples from Europe's regional seas. *Ecology and Society* 19(4): 46.

Pilkshs, M. Kalejs, M. Grauman, G. 1993. Influence of environmental conditions and spawning stock size on the year-class strength of eastern Baltic Cod. *ICES* 22: 1–13.

Röckstrom, J., Steffen, W., Noone, K., Persson, A., Chapin III, F.S., Lambin, E.F., Lenton, T.M., Scheffer, M., Folke, C., Schellnhuber, H.J., Nykvist, B., deWit, C.A, Hughes, T., van der Leeuw, S., Rodhe, H., Sorlin, S., Snyder, P.K., Costanza, R., Svedin, U., Falkenmark, M., Karlber, L., Correll, R.W., Fabry, V.J., Hansen, J., Walker, N., Liverman, D., Richardson, K., Crutzen, P., and Foley, J.A. 2011. A safe operating space for humanity. *Nature* 476: 472–475.

Vahtera, E., Conley, D.J., Gunstafsson, B.G., Kuosa, H., Pitkanen, H., Savchuk, O.P., Tamminen, T., Viitasalo, M., Voss, M., Vasmund, N., and Wulff, F. 2007. Internal ecosystem feedbacks enhance nitrogen fixing cyanobacteria blooms and complicate management in the Baltic Sea. *Ambio* 36: 1–10.

Vitousek, P.M., Mooney, H.A., Lubchenco, J., and Melillo, J. 1997. Human Domination of the Earth's Ecosystems. *Science* 277: 494–499.

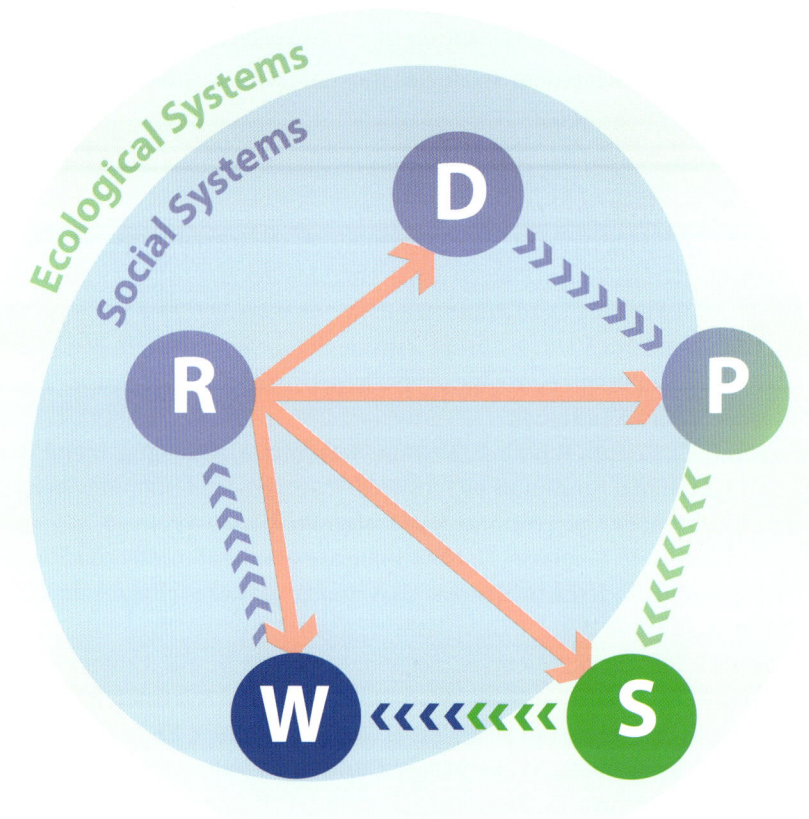

# 3 Let them rule over the fish of the sea and over the birds of the sky

Just as human demands (Drivers and their associated Pressures) on ecosystems change over time, so do human values, expectations from and attitudes toward nature. Differing values result in different relationships between environmental State and human Welfare. In order to best manage our environment these values and the types of Welfare they generate need to be explicitly considered.

On the west coast of North America from about 1820 to 1880, populations of northern elephant seal (*Mirounga angustirostris*), the second largest seal in the world, were heavily exploited for their blubber. These seals bred up and down the west coast from Mexico to California (Fig. 3.1). The species' imperative to breed on land and its large size made sealing an easy and more accessible source of oil than whaling. By 1884 the populations on the American mainland had disappeared and the species was thought to be extinct with the exception of a remote offshore population on Guadalupe Island off the coast of Baja California. In May 1892 an expedition of the Smithsonian Museum discovered a colony of nine individuals at Guadalupe island,

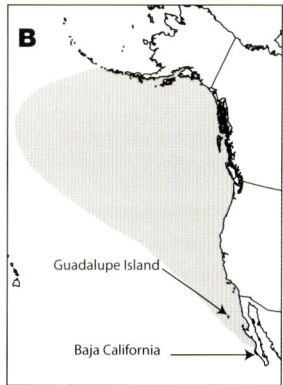

**Figure 3.1** **A**: The northern elephant seal (© Mike Baird) and **B**: its modern habitat distribution (IUCN Redlist data).

and shot seven of them: 'the action was considered justifiable at the time, as the species was considered doomed to extinction…and few if any specimens were to be found in North American museums' (Anthony, 1924). The species was again considered to be extinct, and when small numbers of seals continued to appear on Guadalupe Island further kills were made for collections: 4 in 1904, and 14 out of 40 in 1907. The attitude of the era was that 'this was a severe stroke dealt to a struggling species, but the appetite of science must be satisfied' (Huey, 1930).

Fortunately the northern elephant seal proved to be a very resilient animal and its populations somehow managed to rebound so that current population estimates are in excess of 100,000 (Le Boeuf et al., 1994)[1] though these populations show very limited genetic diversity as a result of the human-inflicted population bottleneck of the late 1800s (Bonnel and Selander, 1974).

Shooting the last remaining specimens of an endangered species is anathema to the sensibilities of any modern ecologist, but attitudes toward the extinction of species have not always been as they are now, and are not always earnest and grave[2]. Our understanding of the extent to which humans can alter the global environment has changed completely over the last century. For example, Thomas Huxley[3] argued at the International Fisheries Exhibition in London (1883) that 'all the great fisheries are inexhaustible'. Today, in light of the declining global fisheries stocks and the series of fisheries collapses, a modern biologist speaking on the topic would be more likely to ask whether any great fishery is sustainable. The history of the last century, the great advances in technology that have allowed humans to proliferate so unprecedentedly across the globe, have concurrently led to the unprecedented losses of species.

If human life since prehistory has been a battle of man versus nature, the industrial revolution may have marked the turning point when man

1 The Stellar's Sea cow, a relative of today's dugongs and manatees, was less fortunate. Discovered in 1741 it was extinct by 1768. Parts of at least 50 can be found in museums around the world representing about 2.5% of the extant population when discovered. http://www.hans-rothauscher.de/steller/museums.htm.

2 See the characterization of the ridiculous Dodo, for example in Lewis Carroll's *Alice in Wonderland* or more recently the Hollywood children's film *Ice Age*.

3 The Huxley family have played a remarkable role in the history of ecological science. Thomas was known as 'Darwin's bulldog' for his defence of evolution when the theory first emerged. *See* note 6 chapter 1 about his grandson Julian.

finally got the upper hand. The strategies evolved by humans that enabled us to survive previous eras may no longer be the most reliable strategies for survival in this one. Extinction rates of animals are now about 100 times greater than the rates of extinctions found in fossil records (Millennium Ecosystem Assessment, 2005). That is, we as a species are killing off 100 times as many other species than other 'natural' processes. Currently about 1.2 million species have been described and classified and catalogued centrally (1.3 million invertebrates; 320,000 species of plants; 60,000 species of vertebrates and 50,000 others) which make up a fraction (about 14%) of the total likely number of species on Earth (Mora et al., 2011). Our capacity to obliterate species is not new. The link between extinctions and human development goes far back in time. The timing of extinctions of large mammals (including mammoths, mastadons, and the like) at the end of the last ice age is closely connected in each continent with the arrival of early humans (Lyons et al., 2004).

The relatively recent recognition of the scale of irreversible human alterations to global ecosystems has led to increasing awareness of the need to protect animals from extinction. Biodiversity, the diversity of life on Earth and the imperative for conservation have been internationally agreed under the United Nations Convention on Biodiversity (UN, 1992).

Biodiversity is a popularly appealing concept. It encompasses the variety of all living things from the molecular to the planetary level. The growth in use of this word since the early 1990s is testament to its great communicative power. However, the act of merging the two words 'biological' and 'diversity' involved reification (thingifying). Referring to biodiversity as a stand-alone concept concretizes the variety of life as separate from (and perhaps more than) its component parts (genes, species, ecosystems). From a practical standpoint, the scientific analysis of biodiversity takes a Cartesian approach. Individual components of diversity are identified (e.g. numbers of species or individuals) and formulae are used to generate indicators describing the mathematical relationships between these components. Despite having metrics to describe biodiversity, there is still a great degree of uncertainty about how biodiversity affects the functioning of ecosystems, and the ecosystem properties such as productivity and stability which it confers, according to theory (Hooper et al., 2005). That is, we can describe biodiversity but we don't really know what it does.

Superficially the idea that biodiversity should be conserved seems trivial. Nobody wants to destroy the variety of life, but in practice there are plenty of aspects of biodiversity that societies can live without, or more importantly, find it difficult to live with. Small-pox, malaria, fleas, ticks, mosquitoes, rats, snakes, bears and wolves: in contact with humans these creatures form a deadly continuum from invisible to charismatic.

While it may not be difficult to make a case for the protection or conservation of larger, iconic species such as the wolf, the tiger, the panda or the polar bear,[4] the protection of smallpox or malaria on the grounds of biodiversity is difficult to defend in practice. Though there is international consensus that biodiversity is important and should be protected, our perception of the aspects of biodiversity that deserve our protection is affected by our own interests and values, and these values are subjective. If the behaviour of the Smithsonian expedition on Guadalupe Island seems strange and wrong to us today with our modern sets of values toward conservation of species, perhaps it is worth analysing our own views toward nature and biodiversity.

Grouse hunting on the upland heather moors of Scotland provides a useful illustration of a relatively well understood ecosystem that elicits strongly contrasting sets of values. Grouse moors have been regulated and managed for hunting for almost two centuries. Aspects of the ecology of grouse management were well understood long before the science of ecology had developed.[5] A brief history of the topic provides a good example of how human values may change over time and how groups with similar aims may hold different values in relation to different parts of the same ecosystem, potentially resulting in conflict.

Scotland for at least two centuries has hosted a major nature tourism industry. Because of the plentiful game, beautiful landscapes and apparently less restricted hunting, sporting (hunting and fishing) holidays for wealthy southern gentlemen starting in the late eighteenth century became hugely popular in the mid to late nineteenth century. The early accounts of hunting expeditions such as that of Colonel Thornton's 'Sporting tour' (1804) helped to popularize Scotland as an untamed wilderness with its favourable

4 Sometimes cynically referred to as a 'charismatic species Olympics'.
5 'There seem to be little doubt that the stock of partridges, grouse, and hares on any large estate depends chiefly on the destruction of vermin,' Charles Darwin in *Origin of Species* (1861).

accounts of the Scottish landscape.[6] In the 1800s the tourist industry burgeoned, assisted first by steamships (in the 1820s) then by the availability of rail travel (in the 1840s). Increasingly large tracts of Scottish land were set aside for the management of game. Rental of land to southern visitors was big business and the most prestigious and most lucrative of the sporting pursuits was grouse shooting, having an estimated value in 1909 of £1.5 million (Durie, 1998) equivalent to about £180 million in modern terms.

Grouse shooting was the preserve of the élite due to the high price of the long travelling distances from London (then, as now, the population and economic heart of the UK). The beginning of the grouse season (12 August) also corresponded conveniently with the holidays of law courts and parliament. In the late nineteenth and early twentieth century it rapidly grew in popularity as a sport, reaching its zenith in the 1930s. Developments in the technology of shooting also made the sport more appealing. The invention of the breach loaded gun (guns with modern cartridges) allowed shooters to point their aim upwards.[7] With this new technology the driven grouse shoot became possible, in which beaters drove the birds toward the waiting shooters, allowing much larger takes of birds with less effort from the shooting participants. The new hunting style also facilitated the less athletic sportsman. King Edward the VII[8] (having lost some of the physicality of his youth) espoused grouse driving, and this royal endorsement of the sport consolidated its position as a highly respectable and desirable pursuit. In the United Kingdom grouse shooting is still associated with a social élite. The industry currently supports around 1000 jobs contributing about £14.5 million in wages in Scotland alone (Fraser Allander Institute, 2010). Figure 3.2 illustrates the grouse bag for the entire UK over the last 150 years. The rise in bags in the nineteenth century is attributable to increased popularity as well as the advent of the driven hunt. The number of birds caught fell to an all-time low for the period during the Second World War. Note that the trend in

6 'South Britons may talk of their beautiful, highly-finished landscapes, of which I have seen the most deserving to be viewed in England, and have been pleased with their elegance and neatness; but from their small pitiful extent, they grow flat and lose their effect. Here the case differs; for the immense extent of these views and the reflection of the sun, presenting various tints, each differing from another, though all beautiful gives this country every advantage and a decided superiority...' Thornton, 1804.

7 Before the invention of the cartridge, guns were muzzle loaded. Pointing a muzzle-loaded gun skyward results in the gunpowder falling out of the gun, and backwards onto the shooter.

8 King of the United Kingdom, the British Dominions and Emperor of India.

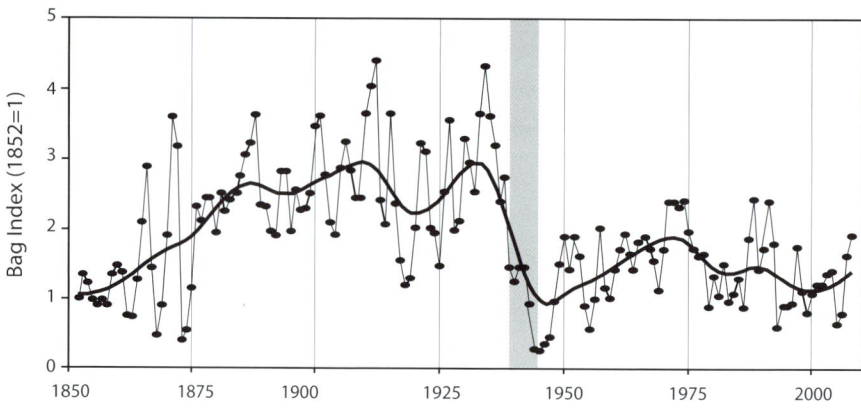

**Figure 3.2** Time series of grouse bag data redrawn from the Game and Wildlife Conservation Trust. http://www.gwct.org.uk/research/long-term-monitoring/national-gamebag-census/data-back-to-darwin/

catch with time is a function of both the amount of shooting (the Pressure) and the size of the grouse stock (State), which varies cyclically with time.[9]

The red grouse *Lagopus lagopus scoticus* (a subspecies of the willow ptarmigan), on which the industry depends, inhabits upland heather moors and feeds on heather, seeds, berries and insects. The long tradition and popularity of grouse shooting has led to the active management of grouse habitat for two centuries. The aim of such moorland management for grouse is to maximize the number of grouse available for shooting. One of the main techniques is called 'strip muirburn': under this practice strips of the heather moorland are periodically burned. The practice encourages the growth of young, tender heather plants, providing an excellent source of food for the birds with the older heather plants in the unburnt strips of moor providing cover and nesting habitat.

Heather moorland is in itself considered an important habitat in terms of biodiversity, as well as acting as a reservoir for carbon storage and source of drinking water (Grant et al., 2012). As a ground nesting bird, the red grouse have many natural predators including birds and mammals. Mammalian predators of grouse include the red fox (*Vulpes vulpes*), mustellids such as polecat (*Mustela putorius)*, pine martens (*Martes martes*) and wild-cat (*Felis silvestris*). Avian predators of grouse include birds of the crow family (Corvidae) the Carrion crow (*Corvus corone*), hooded crow (*Corvus cornix*) and

9 This is exactly analogous to the time series of commercial salmon catch for the Columbia River shown in the Introduction.

the raven as well as many raptors. A major aspect of grouse moor management has been the control of these predators. Predator control, in particular control of the hen harrier (*Circus cyanaeus*), can enhance the productivity of grouse on the moors and the size of the hunters' bag (Thirgood et al., 1995); it has also been demonstrated that protection of the hen harrier can lead to economic unsustainability of grouse shooting estates (Redpath and Thirgood, 1997).

In the interest of grouse populations, these predators have been extensively culled over two centuries by a variety of means. Over the same period many of the mammal and raptor species have declined. Prominent bird species experiencing local extinctions include white-tailed eagles (*Halieaeetus albicilla*), the osprey (*Pandion haliaetus*) and the goshawk (*Accipiter gentilis*), while others such as the red kite (*Milvus milvus*) and golden eagle (*Aquila chrysaetos*) had their populations severely reduced (Harris et al., 1995). In 1954 the Protection of Birds Act preventing the deliberate control of raptors was introduced in the UK. Many of these charismatic species have subsequently undergone programmes of reintroductions. This legal protection, the schemes of reintroductions combined with the decline in organochlorine pesticides and the reduction in the number of gamekeepers following the Second World War, have all contributed to the recovery of many raptor species. However, while these birds are now afforded legal protection, illegal poisoning of raptors still occurs and is more frequent in the vicinity of areas actively managed for grouse (Whitfield et al., 2003).

The protection of birds in the UK has popular support and the Royal Society for the Protection of Birds counts over 1 million members. Considerable economic values are also associated with many charismatic bird species in Scotland. For example, the white-tailed sea eagle was reintroduced to the Island of Mull in 1975. These birds now attract significant numbers of tourists on an annual basis and the value of this tourism is estimated at around £1.5 million per year.[10]

This case illustrates a conflict in objectives for habitat management. The longstanding management has been aimed at the maximization of the grouse bag; this has also provided protection for the heather moorland habitat on which the grouse rely. As such, it has been beneficial to the conservation

---

10 Some sheep farmers continue to object to the presence of the birds due to concerns over the threat of the birds to young lambs.

of what is considered an important habitat for biodiversity (Grant et al., 2012). The management for grouse and the extirpation of predators has also had undesirable consequences in terms of the loss of biodiversity of birds and mammals. The legal protection afforded to bird species in the mid-twentieth century indicates a shift in value toward the conservation of wildlife. The charismatic species favoured by the bird-watching community are exactly those that diminish the stocks of grouse and reduce the quality of grouse shooting.

The history and modern connotation of grouse shooting as an élitist sport has the potential to engender conflict, as the privileged few may be perceived to spoil biodiversity for the bird- and mammal-loving public. Objectively, neither management for game birds nor for raptors is correct. The choice of management objective, whether for game birds or charismatic predators, is entirely dependent on subjective values – whether an individual prefers 'field sports' or nature watching; grouse or birds of prey.

Ecologists and biologists, the experts in biodiversity and ecosystem management, are concerned with empirical facts, reproducible results and the study of the functioning of ecosystems. In the tradition of hard sciences, the tools of ecology are mathematics and the measurement of physical and biological parameters, positive and negative feedbacks and trophic interactions within a food web. People study natural sciences and they become biologists or ecologists for personal reasons, out of a particular interest in nature and the natural environment. Environmental scientists often found their avocation due to a personal interest or love of the environment. While good environmental science is objective, the values driving the research directions of individuals may not always be neutral.

Scientific attitudes towards introduced species provide an example where subjective values toward biodiversity have crept into the mainstream of scientific research. In tandem with the rise in prominence of the term 'biodiversity' and following some disastrous accidental species translocations that occurred in the 1980s, the birth of invasion biology as a field of ecology in its own right reflected a growing concern with the introduction of species to new locations in the 1990s (Davis et al., 2011).[11] Prominent examples include the introduction of the filter feeding zebra mussel

---

11 Though the first major synthesis *The Ecology of Invasions by Animals and Plants* (Elton, 1958) was published forty years earlier.

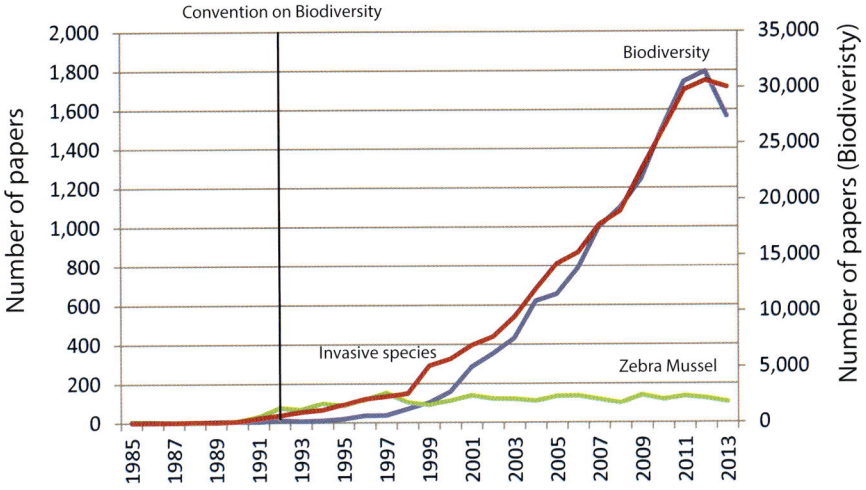

**Figure 3.3** Number of search results for the terms zebra mussel, invasive species and biodiversity in ISI Web of Science 1985–2013.

*Dreissena polymorpha* to the US[12] and the arrival of the comb jelly *Mnemiopsis leydii* in the Black Sea.[13] Figure 3.3 illustrates the relatively recent growth in levels of *research* into introduced species. Until the late 1980s papers about invasive species were uncommon. Between the publication of the classic text on the subject, Invasion Biology (Elton, 1958) and the year 1980, the ISI web of knowledge counts only four papers with the phrase 'invasive species'. In the 1990s growing concern about the arrival of the zebra mussel in the US saw a rise in the number of publications both on invasive species and on zebra mussels (the annual number of papers emerging on these topics was correlated from the mid-1980s to mid-1990s). Gradually, once the zebra mussels had become established (the problem was understood) the number of papers on the zebra mussel began to stabilize;

12 Native to Black and Caspian seas the zebra mussel spread across Europe in the 1800s and was introduced to North America in the mid-1980s. The mussel has the capacity to alter ecosystems through change in phytoplankton and submerged plant composition, competition with native mussels, modify nutrient flows and alter energy flow within ecosystems (Naddafi et al., 2011). *Dreissena* has also caused major economic impacts, particularly in the U.S., through disruption of drinking water treatment and power generation facilities (Connelly et al., 2007).

13 *Mnemiopsis* caused huge economic losses to fisheries in the Black Sea (Knowler, 2005), but the changes to the ecosystem caused by *Mnemiopsis* were mitigated by the subsequent arrival of a second introduced comb jelly, *Beroe ovata*, which acted to suppress the populations on *Mnemiopsis* (Shiganova et al., 2004).

however, the number of papers published on invasive species has grown steadily ever since, and is positively correlated ($r^2$=0.989) with the number of publications on biodiversity. The emergence of these two interrelated fields was coincident with the signing of the convention on biodiversity, and both have been at the vanguard of ecological science for the last 20 years. However, scientific literature describing introduced species has been criticized for adopting alarmist language and military metaphors describing aliens, invaders and invasions. This alarmism has resonated with the public – the idea of pure, natural biodiversity as good, under threat by the non-indigenous invader has proven a popularly appealing narrative.

While the arrival of species to new locations may be unwanted, even disastrous as in the cases above, the appealing narrative of invasion and the subjective view that human introductions necessarily have negative impacts on biodiversity are not always justified (Davis et al., 2011). It is important to recognize that the emergence of concern over non-indigenous species has represented a value shift away from practices that have enabled the growth and spread of human society around the globe.

A list entitled '100 of the world's worst invasive alien species' (Lowe et al., 2000) under its entry on mammals includes the goat (*Capra hircus*) and feral pigs (*Sus scrofa*). According to the list the species were chosen based on their 'serious impact on biological diversity and/or human activities, and their illustration of important issues of biological invasion'. Yet the domestication of animals marked a critical point in the evolution of human societies; both goats and pigs were some of the first animals to be domesticated around 10,000 years ago (Naderi et al., 2008) and they have been deliberately transported around the world with humans ever since. While there is no doubt these animals have damaged biodiversity, even caused extinctions in places, they have also provided immeasurable human benefits since prehistory. Labelling animals and plants that make up integral parts of our social-ecological systems and have enabled human development for millennia as 'invaders', based on new and poorly understood arguments of biodiversity (and biological purity), indicates a level of subjectivity that risks undermining the science of ecology.[14] When making

---

14 This is not to suggest that biodiversity is not beautiful and fascinating or should not receive academic attention, only that it does not necessarily provide suitable objective targets for managing social ecological systems.

environmental management decisions it is important to know what are facts and what are opinions.

Managing the natural environment involves choices. Choosing one set of objectives over another involves decisions about which objectives and targets should be set. Unfortunately for the manager, different sets of users of a particular ecosystem can have directly opposing and equally valid environmental objectives for the same ecosystem, as we saw in the case of the grouse habitat in Scotland.

Any number of similar value conflicts around the world could be used to illustrate the point made in this chapter – from the case of salmon and sea lions in chapter one to the Japanese values towards 'scientific' whaling in the Antarctic (International Court of Justice, 2014) to the constitutionally enshrined religious values towards cattle in the Indian subcontinent (Simoons, 1973), conflicts in values toward biodiversity abound.

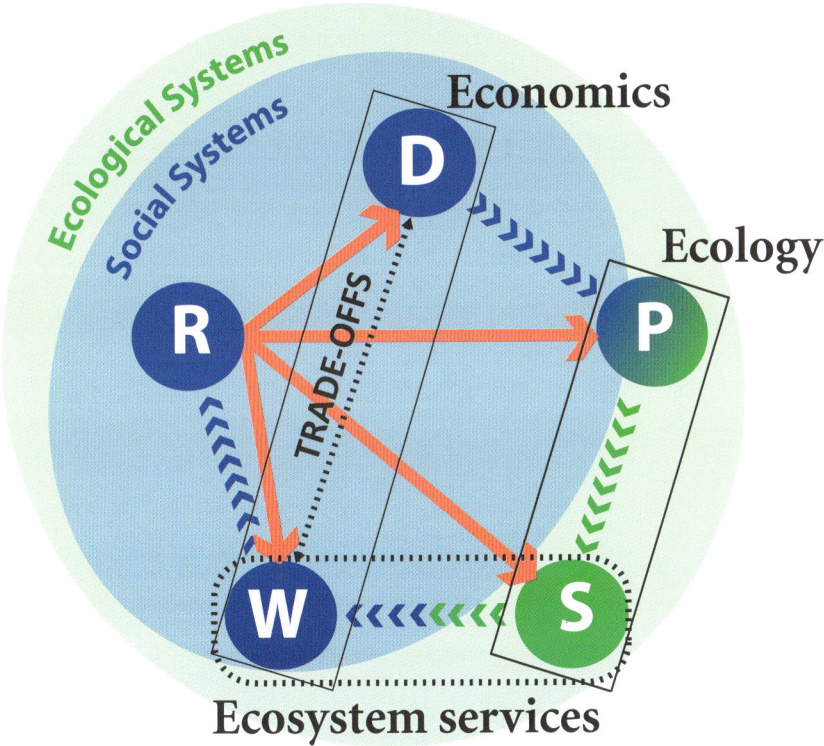

**Figure 3.4** DPSWR diagram showing trade-offs between Drivers and Welfare and indicating the role of ecosystem services.

Returning to the original conceptual framework (the DPSWR) presented in chapter one, the link between changes in environmental State (of which biodiversity is one aspect) and how these affect society lie at the boundary of the social and ecological systems. In Figure 3.4 above, the sphere of expertise of the ecologist lies within the ecological system and is concerned with measuring, understanding, modelling, and predicting the complex interactions that affect the links between Pressures and State changes. The measurement of Welfare is the traditional realm of the economist concerned with human values. Understanding the links between State change and Welfare is the subject of the emerging discipline of ecological economics: the field of ecosystem services and the topic of the next chapter.

## Suggested reading

Davis, M.A., Chew, M.J., Hobbs, R.J., Lugo, A.E., Ewel, J.J., Vermeij, G.J., Brown, J.H., Rosenzweig, M.L., Gardener, M.R., Carroll, S.P., Thompson, K., Pickett, S.T.A., Stroberg, J.C., Tredici, P.D., Suding, K.N., Herenfeld, J.G., Grime, J.P., Mascaro, J., and Briggs, J.C. 2011. Don't judge species on their origins. *Nature* 474: 153–154.

Hooper, D.U., Chapin III, F.S., Ewel, J.J., Hector, A., Inchausti, P., Lavorel, S., Lawton, J.H., Lodge, D.M., Loreau, M., Naeem, S., Schmid, B., Setälä, H., Symstad, A.J., Vandermeer, J., and Wardle, D.A. 2005. *Effects of biodiversity on ecosystem functioning: a consensus of current knowledge. Ecological Monographs* 75: 3–35.

Mora, C., Tittensor, D.P., Adl, S., Simpson, A.G.B., and Worm, B. 2011. How many species are there on Earth and in the ocean? *PLoS Biol* 9: e1001127. doi:10.1371/journal.pbio.1001127. 137pp.

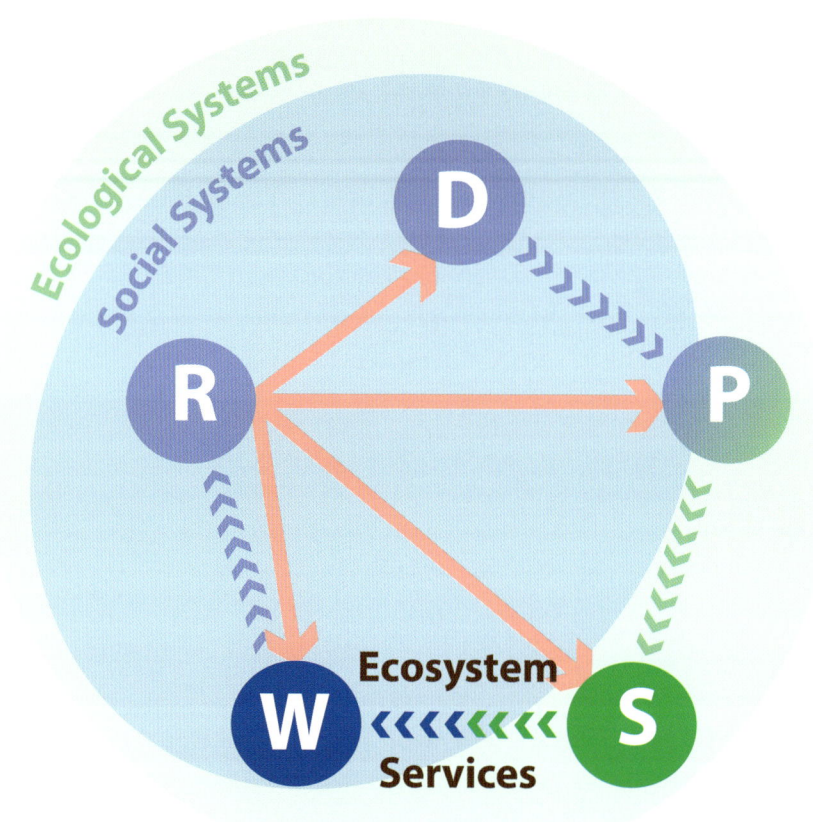

# 4 The poor man wants the oyster, the rich man wants the pearl

**Ecosystem services** are the benefits humans obtain from ecosystems, the aspects of nature used to produce human well-being; these include clean air and water, food and fuel. Humans, as individuals and as societies, are utterly dependent on many of the physical components and ecological processes that make up ecosystems.[1] In the developed world, where most people live in cities at remove from nature, it is easy to take this dependence for granted. Since ecosystems provide us with so many benefits in many different ways, a classification system for ecosystem services provides a useful starting point to understand these connections. The classification scheme presented below (in bold) is that of the Millennium Ecosystem Assessment (2005).

Some of the human benefits from ecosystems are obvious – the sea gives us fish, the pig gives us meat, the tree gives us wood. Products obtained from ecosystems are called **provisioning services** (for example, the cod and the salmon discussed in previous chapters). Other services are less immediately apparent; for example, ecosystems regulate the concentrations of gases in the atmosphere and control climate, they absorb waste and provide clean water. These benefits obtained from the regulation of ecosystem processes are called **regulating services**.

Humans also derive non-material benefits from ecosystems: the thrill of a struggling fish; the sight of an eagle; a vista of the sea at sunset or a forest canopy illuminated by campfire under a sky of shimmering stars – experiences of nature's majesty enrich us culturally and spiritually. Even the idea of existence of a unique ecosystem or an undiscovered species (the Amazon rainforest or the Phyllophora field in chapter 2, for example) may give us a

---

1 Chapter 2 dealt with Pressures, negative effects that social systems place on ecological systems; ecosystems services are the opposite, positive effects flowing from ecosystems to social systems.

warm glow. These non-material benefits provided by ecosystems are collectively called **cultural services**.

Underpinning all of these categories of benefits are the **supporting services** necessary for the production of all other ecosystems services; these include all the critical life-support essential for human societies to flourish, as well as the provision of habitat for animals.

The concept of ecosystem services is a profoundly anthropocentric one. From this perspective, ecosystems are perceived as a means of achieving human Welfare. A fully functioning, perfectly healthy ecosystem would provide no ecosystem services if there were no humans to experience the benefits, and any aspect of an ecosystem only becomes a service when it provides benefits to humans. Economics has long focused on the measurement of costs and benefits of things that are traded on markets, including some of the products of nature (e.g. fish, timber) but only relatively recently has the idea of incorporating the whole suite of ecosystem services into the calculation of costs and benefits begun to be widely adopted and the concepts of ecosystem service refined for practical use.

In the DPSWR the logic of ecosystem service valuation is clear. **Immediate Drivers** of ecosystem change (the farming, hydro-electric development and grouse hunting of the previous chapters) are economic sectors, activities or processes intended to enhance human welfare (to generate benefits). In addition to the economic benefits these Drivers create, they also result in environmental costs (negative changes in human Welfare) through changes in ecosystem State. The destruction of fish stocks (e.g. salmon in the Columbia River and cod in the Baltic), entire ecosystems (the Zernov's Phyllophora field), or aspects of biodiversity (the birds of prey and predatory mammals of Scotland) and these State changes have all resulted in losses to human Welfare of different types, some of which (for example, the fish) are captured in markets while others (the loss of biodiversity) are economic **externalities**. For example, the costs to recreational salmon fishery in the Columbia River are not measured in markets, nor is the biodiversity of the Zernov's Phyllophora field. Understanding (and quantifying) the trade-offs between the costs and the benefits of specific Drivers[2] can therefore help incorporate the ecological parts of the social-ecological systems into management and guide

---

2 A detailed account of the theory and practice of environmental economics is beyond the scope of this book. Field and Field (2002) provide a useful introduction to the topic.

the best courses of actions for development. Ecosystem services offer the potential to make these costs explicit, to quantify them and to incorporate them into environmental management. Figure 4.1 highlights the trade-off between benefits of Drivers and the Welfare costs they can generate through Pressures and changes in environmental State.

The concept of ecosystem services has been around for over a decade (Daily, 1997; Costanza, 1997) and the first major study on global supply of ecosystem services came in 2003 with the Millennium Ecosystem Assessment. This first systematic global assessment of ecosystem services produced sobering findings. The study found that ecosystems have been altered more widely and rapidly to meet human needs in the last 50 years than in any other comparable period. It also found that the gains in human welfare caused by the Drivers of change over the past 50 years have had growing costs in terms of ecosystem service degradation and are exacerbating poverty for many, acting as a barrier to achievement of the Millennium development goals.[3]

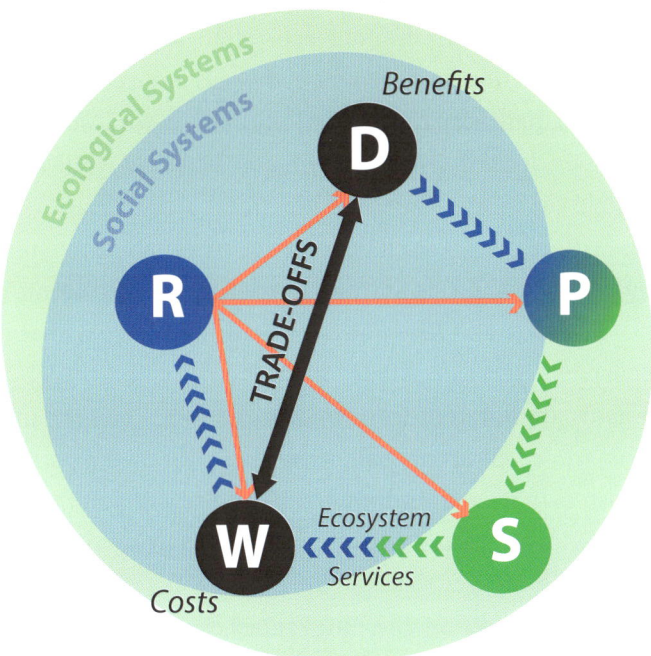

**Figure 4.1** Trade-offs between Drivers and Welfare.

3 To eradicate extreme poverty and hunger ; to achieve universal primary education; to promote gender equality and empower women; to reduce child mortality; to improve maternal health; to combat HIV/AIDS, malaria, and other diseases; to ensure environmental sustainability; to develop a global partnership for development.

The flurry of interest in the application of ecosystem services in environmental management, following the Millennium Ecosystem Assessment, has led to the mainstreaming of ecosystem services concepts and an increasing emphasis on ecosystem services as cash commodities. The study also highlighted the importance and usefulness of the ecosystem services framework as a powerful tool for environmental communication and management, but also revealed the need for further refinement of the ecosystem services concepts if they are to be of practical use.

Subsequently attempts have been made to develop and refine definitions of ecosystem services. These include, for example, the concept of final ecosystem services (Boyd and Banzhaf, 2007), defined as *component of nature directly enjoyed, consumed or used to yield human well-being* for the purposes of estimating Green GDP. Fisher et al. (2009) developed this idea for use in decision making, defining ecosystem services as *the aspects of nature used directly or indirectly to yield human well-being* and distinguishing between **intermediate services** and **final services**.

This system incorporates the four ecosystem service categories of the MEA but the distinction between directly and indirectly used services (final and intermediate services respectively) allows for a solid theoretical framework on which economic valuation may be based while avoiding the pitfall of double counting. Figure 4.2 illustrates a classification of ecosystem services within the DPSWR scheme based on the Millennium Ecosystem Assessment categories and incorporating the definition of Fisher et al. (2009). Fisher et al. (2009) also stressed the unusual characteristics of many ecosystem services. Many of the benefits obtained by humans from ecosystems are different from traditional goods bought and sold on markets. Most traded goods are said to be **rival** (if I own a particular item you cannot also own it) and **excludable** (I can prevent you from using the item) but this is not the case with many ecosystem services, which may be **toll goods** which are not rival but are excludable or **pure public goods** which are rival but not excludable. One crucial characteristic of ecosystems services is that they are **jointly produced**, which is that unlike many man-made systems, one ecosystem or ecosystem component can provide many ecosystem services at the same time (Fisher et al., 2009). In order to understand the complexity of these concepts it is worth considering a simplified example in detail.

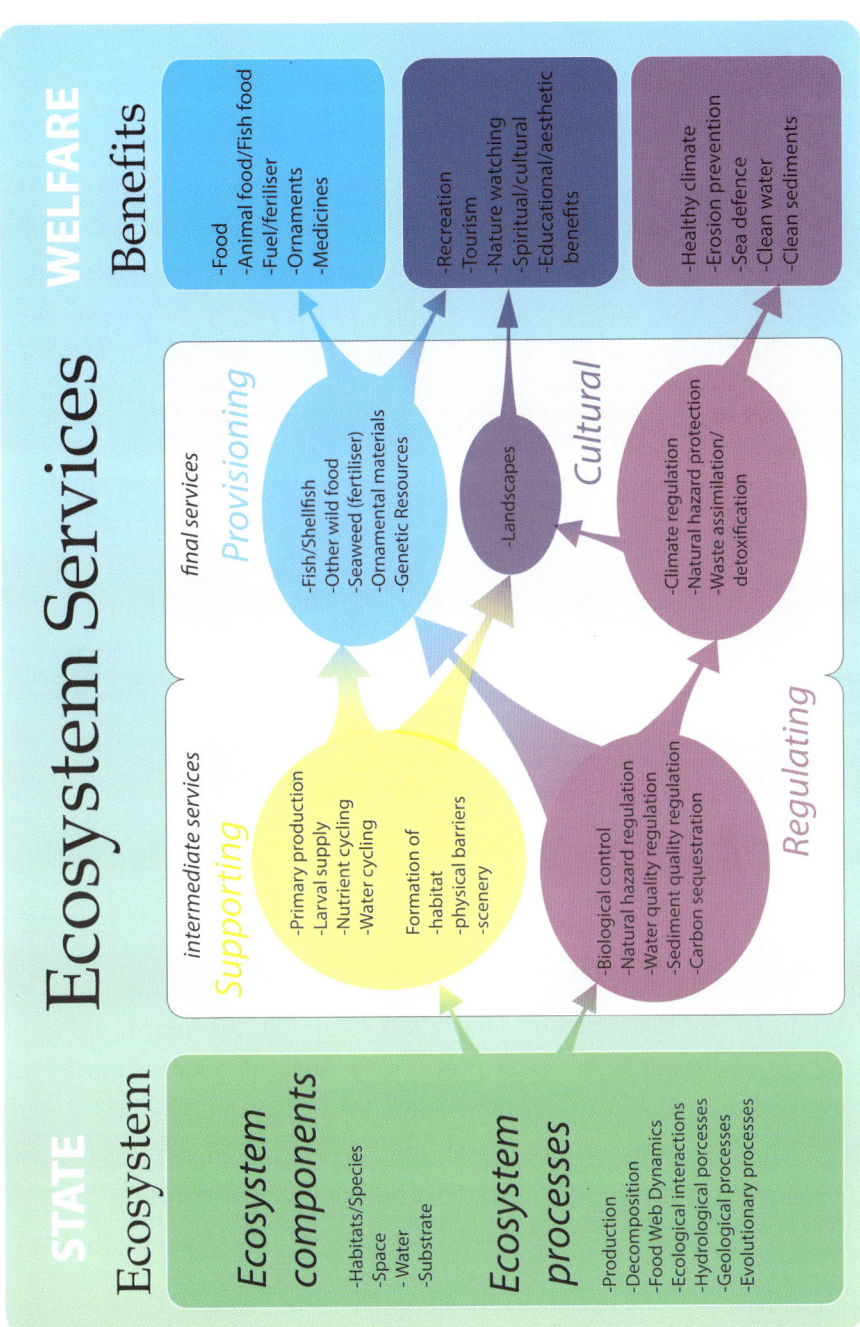

**Figure 4.2** Classification of ecosystem services based on Fisher et al. (2009) and the Millennium Ecosystem classification.

A single tree can provide ecosystem services (Figure 4.3). Perhaps the most obvious of these is the supply of timber, for building or for fuel. This is a **provisioning service**; the wood has a market value and it can be bought or sold. Use of the wood is both rival and excludable – that is, if I chop down the tree, you cannot chop it down (rival), and I can prevent you from using it (excludable).

The same tree left intact can provide many other services. Through the ecosystem process of photosynthesis the tree sequesters carbon, which on the global scale (along with all the other plants) regulates the concentration of carbon dioxide in the atmosphere and affects climate at the planetary scale. The flow of this service is neither rival nor excludable; the benefits I receive from the service do not prevent you from receiving these benefits and I cannot keep these benefits all for myself. The tree also regulates climate at the local scale, providing cool shade in the heat of the midday sun. This service is excludable. If I lie under the tree there may not be space for you, but it is not rival, in that my use of the shade today does not prevent you from using the shade tomorrow. This service, however, depends on whether or not the sun is shining and is only excludable if there are few trees relative to the number of people seeking shade.

The tree may also provide non-material benefits (cultural ecosystem services). I may enjoy climbing the tree for recreation. With the addition of some rope (complementary capital) I can enjoy swinging from the tree. The tree may also provide recreational benefits indirectly (intermediate services) through habitat provision for a favoured bird, mammal or butterfly species. Or it may provide cultural services directly – the tree itself may be beautiful to look at, someone may enjoy drawing or painting it, or it may be useful for teaching us how trees function. Our values toward specific trees or species of tree may be shaped by our history with trees, our childhood memories (or other benefits that may be situation and individual specific)[4].

If I chose to chop down the tree for use as a provisioning service, to sell as timber or to use the land for some other purpose, I should consider not only the benefit I will gain from the use of the tree as timber but also the many

---

4 'There is a tree grows near my house; It's been there quite some time,
   Now, the tree is a slippery elm tree and awful hard to climb.
   But when my wife is after me, in that tree I always roost;
   Why I can go up it just like a healthy squirrel, I don't never need no boost.'
   From *Woodman spare that tree* – Phil Harris.

Aesthetic value [3]

Habitat value [1]

Food [2]

Educational value [3]

Soil formation [1]

Shade regulates local temperature [4]

Water storage [4]

Nutrient cycling [4]

Carbon sequestration and oxygen production through photosynthesis Regulates climate [4]

Recreational use with the addition of complimentary capital [3]

Wood as fuel or construction material [2]

Erosion protection [4]

1. Supporting services     2. Provisioning services     3. Cultural Services     4. Regulating Services

**Figure 4.3** Illustration of the classification of ecosystem services using a single tree as an example.

51

costs I will incur in terms of the other ecosystem services the tree provides, the loss of shade, recreational or aesthetic values. These costs foregone by choosing one option over another are known as **opportunity costs**.

To assess whether using the tree for fuel (the Driver) is a good idea, it is sensible to examine the costs (change in Welfare) of cutting down the tree (the State change). However, only a few of the services described for trees above are bought and sold (wood for fuel or construction, apples for food) and the remainder of the services do not have market values. One approach to valuation would be to calculate the costs of replacing all the goods and services that the tree provides. Steel or plastic might substitute the provisioning service of the wood for construction. Construction of a carbon capture and storage plant might substitute for the regulating ecosystem service of carbon storage, a parasol could be manufactured to replace the shade, a playground to substitute for the active recreational benefits, nesting boxes could account for the bird habitat services of the tree and the aesthetic and cultural values might be replaced by a garden, or a park or a painting.

By adding up the costs of replacing all the ecosystem services provided by the tree, it is possible to make a balanced decision on whether the tree has more value as firewood or if it is better left standing. The assumption here is that adequate substitutes for all the services provided by the tree are available and can be bought; this type of economic valuation relies on the idea that the services of the tree are **fungible**; that is, they can be replaced. However, this is not always the case. For example, efficient technologies for carbon capture and storage have not yet been developed. The assumption that carbon capture and storage technologies might one day replace the carbon storage value of trees is an example of **technological optimism**.[5] The assumptions of fungibility and the resulting technological optimism inherent in economic studies are two major criticisms of the valuation of ecosystem services from a conservation perspective.

The urgency of the requirement for the firewood will dictate the value placed by individuals on the Driver. If I am freezing to death, my immediate need for firewood will outweigh all the regulating and cultural services, the recreational and aesthetic benefits in the short term. I discount the costs in the future for the benefits in the present. While this **discount rate**

5 By contrast the calculation for the Ecological Footprint (Chapter 1) is based on current levels of technology.

is an observed and measurable phenomenon in human behaviour and can be incorporated into cost-benefit analysis, it implies that we place more importance on the present than on Welfare in the future, including the Welfare of future generations, and is often criticized from the perspective of sustainability.

The example of **replacement costs** used above is one of many non-market valuation techniques. Other techniques include **observed preference** methods based on people's behaviour, such as the Travel Cost Method. Here the cost to individuals in terms of time and money spent on travel to climb or swing on our tree can be used to estimate its (non-marketed) economic value.

**Stated preference methods** are another suite of techniques. Individuals could be surveyed to find out how much they would be Willing To Pay to prevent the tree from being chopped down (**WTP value**). Examples of these methods include choice experiments and Contingent Valuation techniques. Detailed treatments of these methods are beyond the scope of this chapter; Champ, Boyle and Brown (2003) provide a useful primer.

Figure 4.4 illustrates some values from a meta-analysis (a statistical analysis of study characteristics and results) from some non-market valuation studies of cultural values for some aspects of biodiversity. While it is possible to economically quantify values for each of the animals in Figure 4.4, from a conservationist perspective the existence value of ten pandas will not substitute for one humpback whale. From an ecological perspective, a humpback whale is not equivalent to Chinook salmon. The WTP value reflects cultural and social values of the individuals responding to a particular study for a particular type of animal, regardless of their knowledge or understanding of the ecological roles of these animals. These values occur in people's minds, and some may have little to do with the biological reality of the State of the environment. Aggregate values for such WTP values may depend more on the number of people holding the values than on the number of pandas or whales in the wild.[6]

The emphasis up to this point has been on the classification and subsequent valuation of ecosystem services. The valuation of ecosystem services has clear uses in terms of cost-benefit analysis of environmental problems

6 Even Bigfoot may have an existence value if people believe it exists.

**Figure 4.4** Some WTP values for different animals. Data from Lopez et al., 2008.

Red Squirrel

Elk

Giant Panda

Brown hare

$2.87

$206.93

$13.81

$0.00

Humpback whale

Bald Eagle

Atlantic Salmon

Chinook Salmon

$128.34

$114.67

$9.45

$126.66

– it can provide an estimate of the costs of economic externalities, helping to assess the importance of a particular environmental problem. In the logic of neo-classical economics it has the potential to rectify this **market failure**. In more pragmatic terms it can communicate environmental damage in a currency that is understood by all, and can help inform the allocation of scarce funds for environmental management. It is often argued that in the absence of valuation a default value of zero may be applied to nature. While there is a clear logic to its use in analysing the costs and benefits associated with a particular Driver activity, the idea of valuation of ecosystem services in application to conservation at the global scale is controversial.

The most influential[7] and controversial examples of a study evaluating ecosystem services is that of Costanza et al. (1997). Published in the prestigious journal Nature, this paper attempted to calculate the value of the world's natural capital and ecosystem services; the value of the whole planet (Figure 4.5).

The authors listed and identified different ecosystem services provided by different ecosystem types. By mapping published economic values for each ecosystem service and considering the area of different ecosystems, they arrived at a value for natural capital and ecosystem services of $33 trillion, almost double the world annual GDP at the time.

The paper was criticized on many fronts[8] from the philosophical to methodological. For some the idea of valuation is unpalatable and difficult to accept; some hold that species or ecosystems have intrinsic values[9]. For others, the idea of valuation of ecosystem services represents an expansion of orthodox neoclassical economic ideology and the theoretical limitations that go with it (Chee, 2004).[10] Some have raised concerns over the emphasis on valuation as a means of commodifying nature and subsuming it into the frame of mainstream economic thinking (Gomez-Baggethun and Parez, 2011). Given that there is empirical evidence that markets erode moral values (Falk and Szech, 2013) at a fundamental level the assumptions of neo-classical

7 As of 2014 the paper has been cited over 11,000 times.
8 Though no responses were published in *Nature*, an entire issue of the journal *Ecological Economics* (Volume 25) was dedicated to commentary and criticism.
9 For a thorough description of this debate *see* Sandler, 2012.
10 Neoclassical or mainstream economics has been described as 'autistic' for its '*marked deficits in communication and social interaction, preoccupation… abnormal behaviour, such as repetitive acts and excessive attachment to certain objects*' Alcorn and Solarz (2006).

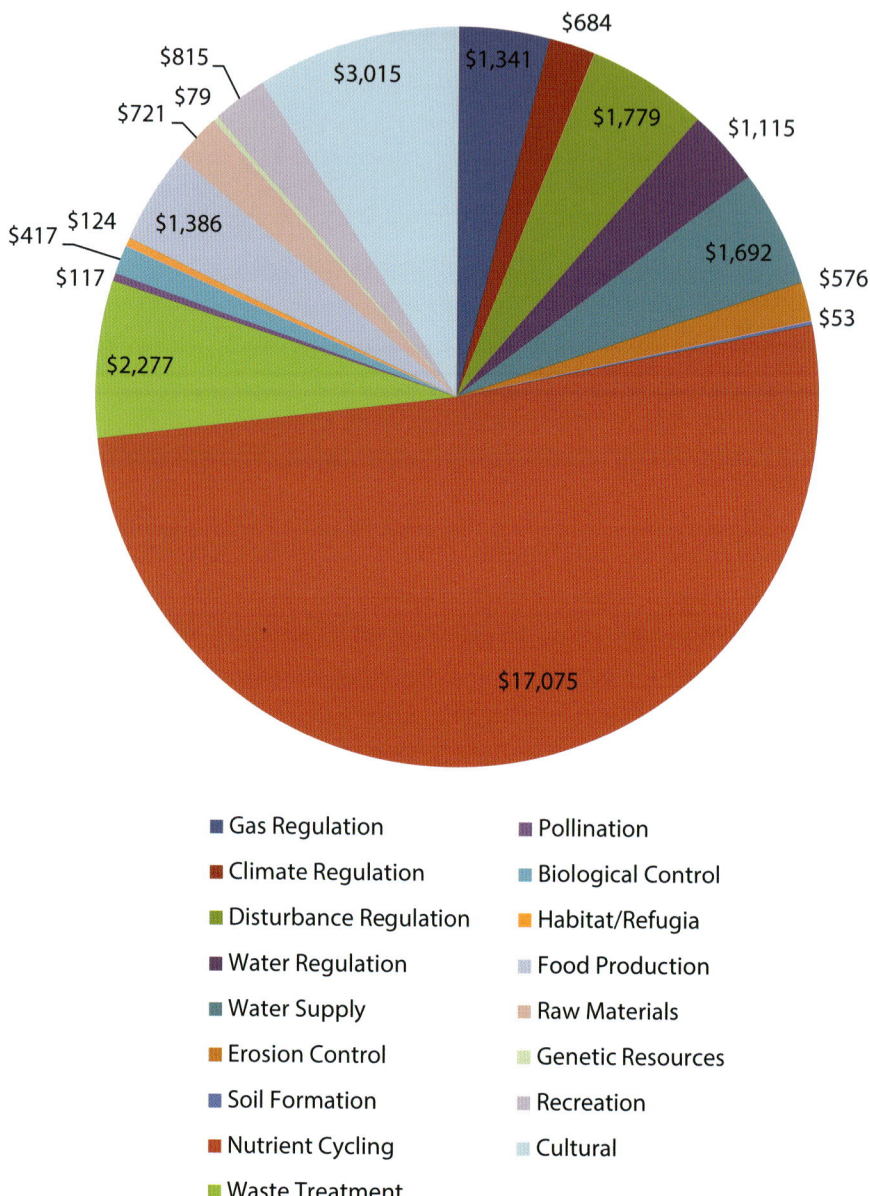

**Figure 4.5** Results of the controversial study (Costanza et al., 1997) estimating the value of the world's natural capital.

economics – that individuals behave out of self-interest as rational actors maximizing their own utility (which works well for private goods bought and sold on markets) – may not be the most appropriate model to adopt for public goods, where equity and sustainability are objectives rather than individual benefits (Costanza and Folke, 1997). Schröter et al. (2014) review some of the most persistent criticisms of both the concepts of ecosystem services and the practice of valuing them.

Despite the criticisms, the potential of the ecosystem service concept to integrate ecological and economic considerations and internalize the externalities of environmental damage fully into societal decision-making has been taken seriously at the international level (MEA 2003; 2005).[11] A recent update to the original Costanza paper has been published (Costanza et al., 2014) exploring how ecosystems service values have declined since 1997 and addressing some of the criticisms of the original paper.

The properties of ecosystem goods and services are highly variable, including continuums of rivalness, excludability and fungibility. These properties make them different from those commonly bought and sold in markets, and therefore more difficult to value economically. While some people object to the concept of non-market valuation for ecosystem services, there are many methods for social decision-making that don't necessarily require the use of economic valuation but may usefully incorporate the unique properties of ecosystem services to achieve environmental management solutions based on a fuller picture of such social-ecological characteristics.

Ecosystem services generally have a spatial dimension; that is, the service is produced in a place. Understanding the spatial distribution of the human values relative to the place can be a key determinant of the overall value (whether monetarily quantified or simply ranked) (O'Higgins et al., 2010; Jordan et al., 2012). These spatial properties are beginning to be used in participatory spatial **Multi-Criteria Analysis (MCA)** for environmental decision-making. MCA can incorporate many goals and objectives. Such objectives may be set around the conservation or preservation of particular suites of ecosystem services. Some groups, for example fishermen or farmers, may want to ensure the continuation of provisioning services in a particular

---

11 With the signature of the Malawi principles, Ecosystem services became the new buzzword in environmental science. Current national or international calls for scientific proposals are littered with the language of ecosystem services.

landscape or seascape, while conservation interests might be the protection of biodiversity and other groups may favour the aesthetic views or a particular landscape. By formalizing such objectives in any particular site and getting spatially explicit values for each criteria, group decisions incorporating trade-offs between multiple ecosystem service considerations can be reached (Arciniegas, 2011; Alexander et al., 2012; Klain and Chan, 2012).

Ultimately all human activities are entirely reliant on ecosystems, and to humans the value of global ecosystems is infinite. Yet changes in ecosystems cause changes in human welfare, and when problems are framed appropriately, ecosystem services can be a useful way of examining these changes. Consideration of ecosystem services can help us to make better and more informed decisions about how to manage the ecosystem in which we live, and valuation can make explicit the trade-offs between Drivers, and the externalities they result in.

However, priorities and values depend on standards of living. My perspective on biodiversity as a middle-class middle-aged white male with almost a decade of university education is not the same as those of the three billion individuals living in poverty. Just as the WTP for recreation in an affluent area is much higher than the recreational values in places that are less well off (O'Higgins et al., 2010) so the WTP for preservation of the panda (Figure 4.4) is almost two weeks of income for somebody trying to live off $1 per day. Under conditions of extreme hunger few would give a thought to the plight of the last Panda, blue whale or blue-fin tuna. Immediate concerns over personal well-being will always trump long-term concerns over conservation so long as poverty persists. For most, urgent concerns for survival take priority over the lofty ideals of conservation, aesthetic and existence values. The value of biodiversity and conservation can only be appreciated in the short term when societies are sufficiently wealthy to realize these values. In order to conserve ecosystem services and the benefits they provide for us, more equitable societies need to develop. Ecosystem service concepts certainly have a role to play in this process. To achieve equity, new forms of management of ecosystem services will need to go beyond the self-interested market-driven neo-classical economic perspective by recognizing and incorporating the unique properties of ecosystem services into management.

## Suggested Reading

Chee, Y.E. 2004. An ecological perspective on the valuation of ecosystem services. *Biological Conservation* 120: 549–565.

Costanza, R., d'Arge, R., de Groot, R., Farber, S., Grasso, M., Hannon, B., Limburg, K., Naeem, S., Paruelo, J., Raskin, R.G., Sutton, P., and van den Belt, M. 1997. The value of the world's ecosystem services and natural capital. *Nature* 387: 253–60.

Falk, A. and Szech, N. 2013. Morals and Markets. *Science* 340: 707–711.

Fisher, B., Turner, K., and Morling, P. 2009. Defining and classifying ecosystem services for decision making. *Ecological Economics* 66: 643–653.

Gomez-Baggethun, E. and Ruiz Parez, M. 2011. Economic valuation and commodification of ecosystem services. *Progress in Physical Geography* 35: 613–628.

Schröter, M., van der Zanden, E.H., van Oudenhoven, A.P.E., Remme, R.P., Serna-Chavez, H.M., deGrooet, R.S., and Opdam, P. 2014. Ecosystem services as a contested concept: a synthesis of critique and counter-arguments. *Conservation Letters* (online) 1–10.

### Websites

http://www.maweb.org/en/index.aspx
http://www.ecosystemvaluation.org/

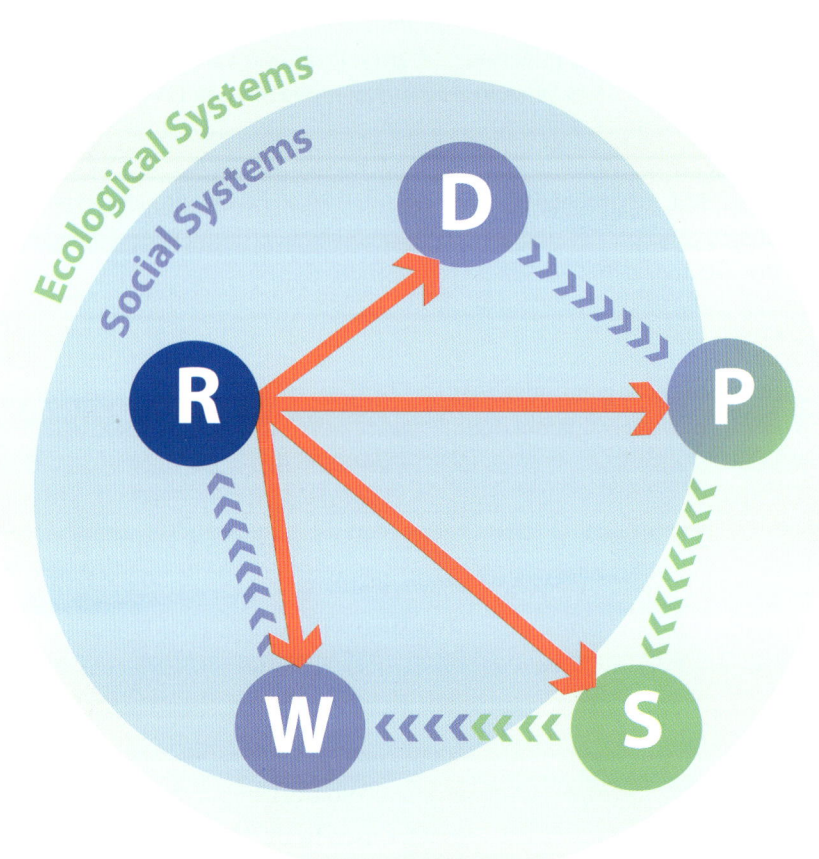

# 5 Targeting actions: Options for Response

Faced with a particular environmental problem and its effects in terms of human Welfare, we as a society may wish to act to solve this problem; good environmental management involves choosing the appropriate Response. Examining the trade-offs between the benefits of a Driver and the environmental costs (the Welfare) through the frame of ecosystems services, we can determine the social importance of the problem[1]. Under the DPSWR scheme the Response can be directed at any of the other elements. The amount of a specific Driver activity may be regulated; technological solutions may be found to change the relationship between the Driver and the Pressure; ecosystems or ecosystem components may be restored directly by acting on State, or the individuals who have borne the costs of environmental degradation may be compensated to restore their lost Welfare. The following sections describe some successful and less successful examples of management Responses to particular problems. These examples do not necessarily represent an ecosystem approach, holistic systems thinking or deliberate targeting of Response at the most appropriate DPSWR element; rather, they are a selection of different management actions that happen to act on particular elements of the social ecological system.

As we have seen in chapter 2, Responses to underlying Drivers such as population or affluence entail very hard choices. Curtailing the growth of population or limiting the levels of affluence of a society can have serious moral, ethical and social consequences and may impinge on human rights. A more practical approach is to regulate the levels of activity of immediate Drivers (economic sectors). Restrictions on the level of an activity are sometimes called **input controls**. These include the introduction of **spatial controls** restricting the spatial extent of the activity (for example, game

---

1  There are many textbooks on cost benefit analysis.

preserves or marine protected areas) as well as **temporal controls**[2] (for example, hunting or fishing seasons) which define specific times during which an activity is permitted.

In some rare cases entire economic sectors have been all but abandoned for conservation reasons. The most prominent example of a complete ban is probably that of whaling. Figure 5.1 shows a twentieth century timeline of Antarctic whaling. Modern whaling[3] came into being in the late nineteenth century with the invention of the exploding harpoon by the Norwegian shipping magnate Svend Føyn. In combination, first with steamships and later (in the twentieth century) diesel ships, the new methods facilitated efficient hunting of the faster whale species (rorquals, baleen whales of the family Balenopteridae) and consequently industrial scale whaling propagated rapidly across the globe. The blue whale (being the largest) was the primary target at first and the Blue Whale Unit[4] became the standard measurement for fixing whaling quotas in the 1930s (by which time the blue whale fisheries had already reached their all-time peak and begun to decline). The next largest species and the next logical target was the fin whale.

Following the Second World War and the temporary cessation of commercial whaling during the war years, the International Whaling Commission (IWC) was founded with the signing by the major whaling nations[5] of the International Whaling Convention in 1946 with the aim 'to provide for the proper conservation of whale stocks and thus make possible the orderly development of the whaling industry'. The convention was an example of a **command and control**, direct regulation of an industry that states what is permitted and what is illegal. It set quotas for individual species. By 1949 the IWC had imposed quotas on humpback whale catches. Quotas and other measures designed to limit the perturbation to the environmental State (in this case size of the whale stock) are termed **output controls**. Subsequent quotas were set sequentially for blue, fin and sei whales, as the catches of

---

2 Consider the example of the rapidly alternating duck and wabbit seasons in the famous Bugs Bunny cartoon *Rabbit Seasoning* (1953).

3 For more than the brief summary here, Tonnessen and Johnsen (1982) do the subject justice in their 800 pages.

4 Blue Whale Unit: 1 Blue whale = 2 Fin Whales = 2.5 Humpback whales = 6 Sei whales; under this scheme, the whales are functionally equivalent and substitutable.

5 Argentina, Australia, Brazil, Canada, Denmark, France, Iceland, Mexico, the Netherlands, Norway, Panama, South Africa, the Soviet Union, the United Kingdom, and the United States.

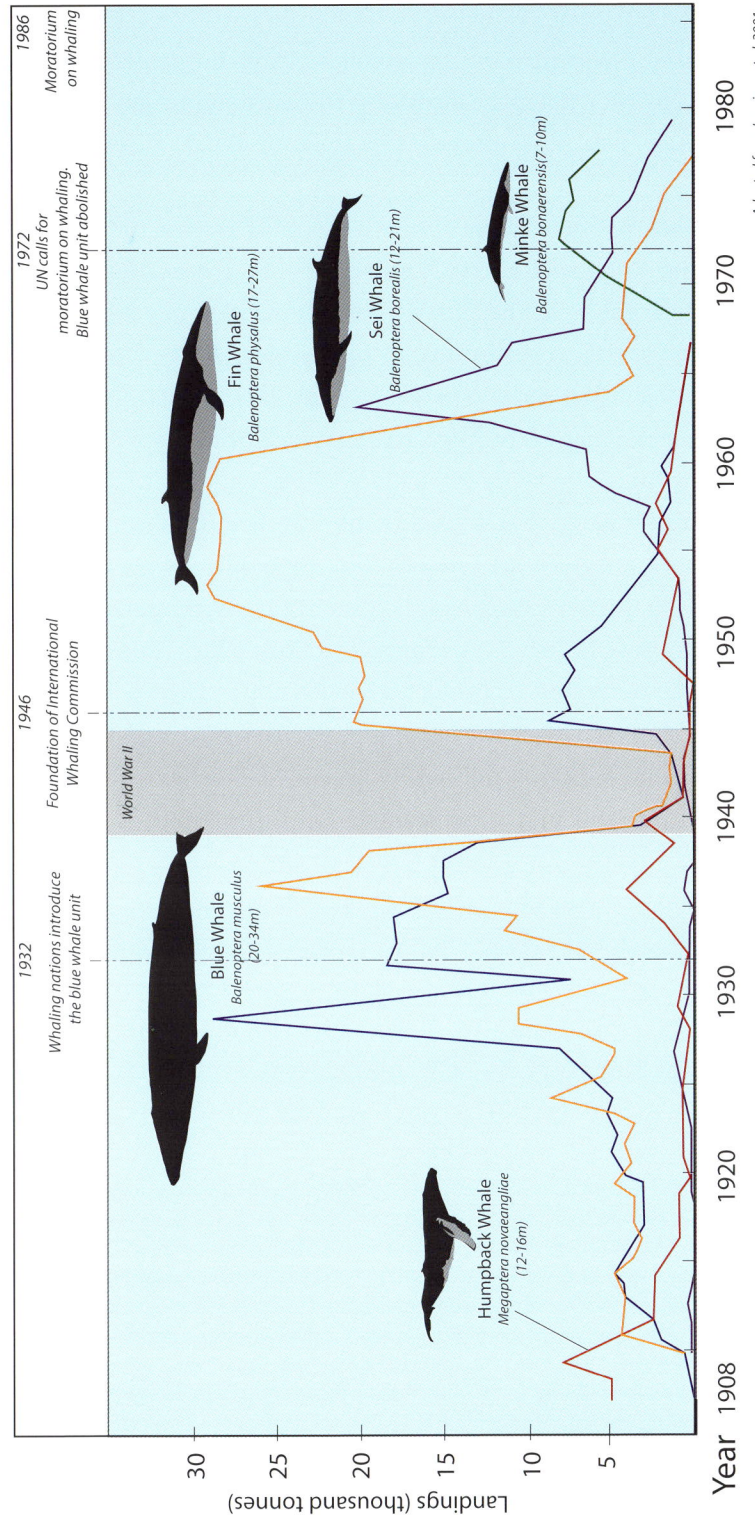

**Figure 5.1** Time series illustrating sequential collapse of the Antarctic whale species. (Re-drawn from Jennings et al, 2001.)

63

these larger target species all began to decline terminally. Lacking enforcement, a pattern of sequential over-exploitation led to the depletion of all the major stocks. In 1972 the UN called for a 10 year moratorium on whaling, but attempts to pass this moratorium through the IWC were initially unsuccessful.

By the 1960s public awareness of whales (and marine life in general) was beginning to grow; the 1963 movie *Flipper*, for example, featured an intelligent dolphin in the Florida Keys who befriends a young boy.[6] Other popular media contributing to public awareness of undersea life included the television series *The Underwater World of Jacques Cousteau* and in 1970 a popular recording of whale vocalization was released for sale under the title *Songs of the Humpback whale*. The anti-whaling actions of Greenpeace, in particular, drew attention to the opposition to whaling and brought whaling as a major issue into the public sphere through the medium of film. Perhaps the pivotal moment for international sentiment about whaling was the 1975 Greenpeace action, which showed activists off California attempting to protect whales from Russian whaling ships. The whaling ship fired over the heads of the activists to harpoon a whale. This reckless and intimidating action provided excellent dramatic and bloody television and may have marked a turning point in public attitudes towards whaling. Kawashima (2005) describes how media portrayals of whales transformed public attitudes towards whales from large fish to almost mythical creatures emblematic of the planet and of conservation.

With growing public and political concern globally, the World Wildlife Fund began encouraging non-whaling nations to join the IWC. The resulting new membership of the IWC effectively gave control of the Commission to nations that did not support whaling. By 1982 a 10-year moratorium on

6 According to the documentary *The girl who talked to dolphins*, produced and directed by Christopher Riley, Dolphins at Marine Studios in Florida which had been used to film *Flipper* were moved to a purpose-built NASA funded research lab in the British Virgin Islands in the early 1960s. Here the dolphins and humans lived together. The dolphins were used as subjects for research into inter-species communication (to support the search for intelligent extra-terrestrial life). One of the dolphins formed a physical and emotional bond with its trainer, Margaret Howe. Little progress was made on the communication project and in frustration, the chief scientist John Lily attempted to improve communication by expanding the consciousness of this dolphin through LSD. The research project was eventually disbanded and the dolphin removed to a smaller indoor facility in Miami with no natural light, where it apparently became despondent and soon died of its own will, by making the conscious decision not to come up for air.

whaling was approved by the members of the commission. By 1986 it was in force. Despite estimates of potentially sustainable catch levels for minke from the IWC scientific committee and the development of a revised management strategy, the IWC has continually upheld the moratorium. A number of major whaling nations did not comply with the moratorium. Norway, having objected to the moratorium at the outset, was free to opt out of it and still maintain a whaling industry focused on the minke whale. Iceland left the IWC following the commission's refusal to accept the recommendations of its scientific council (that a limited whaling of minkes could be performed sustainably) and rejoined it stating an objection to the moratorium, thereby allowing themselves to recommence commercial whaling. The Japanese used Article VIII of the Convention (permitting licensed scientific whaling) as a loophole in the Convention to continue commercial whaling operations. The Japanese government has also tried the same strategy as the WWF encouraging recipients of Japanese aid to join the Commission in support of whaling. Most recently arguments to prevent the Japanese 'scientific whaling' have been focused on the nature of the supposed research, and the International Court of Justice (ICJ, 2014) has ruled that the fleet is not engaged in scientific research and ordered the whaling to be halted.

While the moratorium on whaling might be perceived as a victory for conservation interests, the Driver of whaling and its potential to provide benefits in the long term to some sectors of society (for example, the coastal communities of Norway) has led in part to the foundation of the North Atlantic Marine Mammal Commission (NAMMCO), a pragmatic step offering a potential alternative to the IWC; an adaptive response of whaling nations to pursue rational exploitation. NAMMCO was founded in 1992 following the renewal of the moratorium on whaling. NAMMCO state similar aims to the IWC but is composed of exclusively whaling nations (Norway, Iceland Greenland and the Faeroe islands) and is free to pursue a rational approach to the conservation AND exploitation of marine mammals. Critically, the membership of NAMMCO is open but only on consent of the existing signatories (Article 10). This feature of the agreement can thereby protect the agreement from hijacking by the absolutist non-whaling nations, which now comprise the majority of the IWC (and arguably have perverted its function).

The example of modern whaling illustrates many challenges to the regulation of an entire sector on a global scale. The Convention itself has not met its original objectives: the development of the fishery was not orderly, and most whale stocks have not been conserved at levels high enough to support continued exploitation. **Enforcement** and **compliance** were major problems with the Convention; recommendations made by the Commission were unenforceable, lacking any mechanism to hold members to agreed regulation, and despite the IWC, whaling after the Second World War remained more or less a free-for-all with many nations not complying with the agreed regulations. It was the upsurge in public interest (change in public values over time) and the subsequent political actions of the WWF, using the Convention as a structure for non-whaling nations to transmit the wider public concerns, that achieved the moratorium. The structure and roles of the IWC were not able to adapt to the changing biological and social realities of the whaling issue, to meet their original purpose. The open access clause of the Convention was a weakness (from the whaling perspective) because it allowed nations with a different agenda and set of norms into the negotiation and the resulting division, lack of trust and opposing views have forced pro-whaling nations to work outside of the Convention, or in the case of Japan, misuse the Convention.

The moratorium on whaling also illustrates that banning an entire sector by making it illegal can drive the sector away from regulation – the sector may continue to operate, but outside the scope of the existing law or structure. Economic sectors may often adapt quickly to changing conditions while legislative bodies, with their reliance on procedure and consent, often tend to be less reactive and adaptive.

More generally, the global collapse of whaling illustrates that timescales for political action at the global scale are slow. The speed of political action can often pose difficulties for conservation. There were 14 years between the initial call for a moratorium and the entry into force of that same imperfect moratorium. The whaling example illustrates the difficulties of designing effective institutions at the global scale as well as of managing, extinguishing or eradicating an immediate Driver for purely conservation purposes. Whaling represents a special case in that the goal of conservationists (a complete ban on the killing of whales) is in direct opposition to the aim of the Driver (the harvesting of whales). For most sectors the conservation goals are not so directly opposed to the goals of the Driver itself. For fishing, overfishing and depletion of stocks are undesirable, but people generally do not object to sustainable fishing, and it may be possible through technology, such as improved gear, to change the relationship between the Driver and the Pressure.

The case of nutrients in the Black and Baltic Seas from Chapter 2 provides another example of management efforts, this time at a more local scale. Major efforts have been made in these seas by regional management organizations to alleviate the problems caused by nitrogen and phosphorus. In both seas the main Response has been a programme of development of wastewater treatment infrastructure to change the relationship between the levels of Drivers of wastewater (human effluent) and agriculture and the nutrient Pressures that result from them. In the River Danube catchment area of the Black Sea (the main source of nutrient to the Black Sea), this has been carried out by the International Commission for the Protection of the Danube River (ICPDR), which was founded in 1994 and came into force in 1998. Over €3.7bn was spent by the ICPDR to reduce eutrophication in the period 2000–2005 through investment in improvements to wastewater treatment facilities (ICPDR, 2007). A further €6.2bn is expected to be spent on urban wastewater treatment under the current joint programme of measures, which extends to 2015 (ICPDR, 2012).

Similarly in the Baltic, as understanding of the problem began to develop in the late 1980s, the Convention on the Protection of the Marine Environment of the Baltic Sea Area (Helcom), established in 1974, began to implement measures limiting nutrient pollution. In 1988 a target was set to reduce nutrient inputs to the Baltic by 50% by 1995. Since this time a series of reduction targets has been introduced sequentially. While wastewater treatment has meant that targets for phosphorus have generally been met in both seas (HELCOM, 2009; IDCPR, 2007) targets for nitrogen have been less easy to reach.

In part, the difficulty in lowering the amounts of nitrogen is due to the prevalence of diffuse sources of nitrogen, in particular through agriculture. Modelling studies suggest that in the Danube catchment 86% of nitrogen emissions and 71% of phosphorus emissions now come from diffuse sources (ICPDR, 2009). Similarly for the Baltic, about 70% of nitrogen inputs now leave diffuse sources, of which 80% is attributed to agriculture

(HELCOM, 2009). Measures to reduce the nutrient pollution in the European Union parts of the Baltic and Black Sea catchments include the nitrates directive that regulates the way fertilizer can be applied in agriculture. The spatial distribution of diffuse emissions means that they are more difficult to control than discrete point sources, which can be addressed with end-of-pipe wastewater treatment solutions. However, given the large spatial scale of agriculture and the diffuse nature of the inputs, monitoring, enforcement and compliance of the nitrates directive in the EU are major challenges to achieving nutrient reduction targets. Implementation of the directive needs to be enforced at appropriately fine spatial scales (O'Higgins et al., 2014). Better nutrient management at the scale of farms is required, and while there are relatively simple mechanisms for achieving this – for example, leaving fallow the areas immediately adjacent to streams – putting these methods into practice over the large swathes of Eastern Europe is a major challenge, given that the costs of leaving fallow areas are borne by the farmers but the benefits are borne by those downstream. Solving the problem of agricultural inputs may require farmers to pay for the nutrient pollution they cause (internalization of the externality) at more local scales. It remains to be seen how successful programmes of agricultural nutrient management can be implemented over the whole of Eastern Europe with differing social, economic and cultural conditions.

Despite the great efforts over many years to reduce the Pressures on both the Baltic and Black Sea environments and the effective mass installation of wastewater treatment, further reduction is confounded by the diffuse spatial nature of much of the remaining nutrient pressure.

The Black and Baltic Seas have shown mixed reactions to changing nutrient loads. In the Baltic Sea, while the total nitrogen and phosphorus inputs have been successfully curtailed since the mid-1990s, the total phosphorus concentrations have not decreased accordingly (HELCOM, 2014). This is due to the specific physical conditions in the Baltic and limited flushing of the system through exchange with the North Sea. A pool of accumulated phosphorus in the Baltic sediments is still present and fuels continued eutrophication (Munkes, 2005). The ecological characteristics of the Black Sea are not fully understood, but the sea has been slow to return to its original pre-eutrophic state despite the release of nutrient pressures (Oguz et al., 2012). The unpredictable reaction of the systems to management of ecological conditions in the Black and Baltic Seas mean that management strategies bring uncertainty to whether these strategies for reducing nutrient Pressures will bring about the desired effects in terms of ecosystem State and human Welfare.

For some environmental problems, where stocks have been depleted or particular habitat types have been removed, management efforts may focus on the direct restoration of environmental State. In the Columbia River example in the Introduction, the reduced abundance of fish due to various Drivers has led to huge efforts at restoration of environmental State (size of the fish stock) through the production of hatchery-reared fish. Hatchery-reared salmon have been released annually over the last century to make up for reductions in wild populations. Annually over 300 million salmon from hatcheries are released into the Columbia River system to meet the needs of commercial, recreational and tribal fisheries (Paquet et al., 2011). These hatchery programmes have been successful in ensuring a supply of fish for fisheries. As understanding of the Columbia River salmon develops, management goals have begun to extend beyond the objective of simply maintaining fisheries. Conservation of endangered distinct naturally spawning populations adapted to particular habitat conditions of particular streams and tributaries is now an equally important management objective (Paquet et al., 2011). Naturally spawning populations of salmon are declining, both through the effects of harvesting and through adverse impacts of hatcheries that include competition with and predation of wild stocks. Hatchery-reared salmon are less successful breeders (Williamson et al., 2010) and interbreeding with wild stocks can reduce the overall productivity of the wild spawning stocks (Waples, 1999).

In many cases environmental losses caused by Driver activities have caused direct and irreversible losses in human Welfare that is not readily undone. There are many examples of toxic by-products of Drivers and causing externalities through effects on human health. Under the 'polluter pays' principle the agents that generated the externality may be obliged to compensate those that bore the costs of the externality.

One dramatic and contentious example is that of the lawsuit filed by the Ecuadorian government seeking damages against the oil company Chevron (who bought TexPet, a subsidiary of Texaco) which had drilled for oil in the country for 28 years. Residents of Ecuadorian and Peruvian rainforests alleged that Texaco had...

> ...polluted the rain forests and rivers in Ecuador and Peru during oil exploitation activities in Ecuador between 1964 and 1992. In particular, ... that Texaco improperly dumped large quantities of toxic by-products of the drilling process into the local rivers, contrary to prevailing industry practice of pumping these substances back into the emptied wells ... that Texaco used other improper means of eliminating toxic substances, such as burning them, dumping them directly into landfills, and spreading them on the local dirt roads ... that the Trans-Ecuadoran Pipeline, constructed by Texaco, has leaked large quantities of petroleum into the environment ... they and their families have experienced various physical injuries, including poisoning and the development of pre-cancerous growths.
>
> *Jota v. Texaco (1998)*

The case followed a familiar environmental narrative pitching the impoverished rainforest dwellers against massive global oil interests. Following dismissal from American courts and years of legal proceedings, the case was finally tried in Ecuador.[7] The court found that 60 million cubic metres of oil had been spilt over the period of exploitation, contaminating the water table, surface water, soil, flora and fauna. The Ecuadorian court awarded US$18 billion, including US$8.6 billion in costs and a further $8.6 billion as punitive damages. Naturally the case attracted great attention from the environmental movement internationally, and the narrative of oppressor and victim (which at times went beyond the realm of fact)[8] held popular appeal. However, though an enormous financial settlement was awarded, Chevron successfully sued the plaintiff lawyer for corruption using US Racketeering Influenced and Corrupt Organization Act (RICO) laws[9] and obtained an injunction from US courts to prevent the collection of the damage award. This is an example of an unsuccessful implementation of the polluter pays

---

7 The complexity of the international legal proceedings is beyond the scope of this book.

8 For example, the extinction of one indigenous people, the Sansahuari, was widely reported to have been caused by the activities of oil exploitation. In fact, the tribe never existed at all (Reider and Wasserstrom, 2013).

9 Most famously used against the US Mafia.

principle; in this case (to date) the legitimacy of the court decision has been undermined by the corruption of the process. While the US courts made no findings on the environmental aspects of the Ecuadorian case, they did have legitimate concerns over the legal process, and the legal battle goes on.

Similar cases for compensation have occurred within the US and the US legal system has developed (through unfortunate necessity) a strong basis for the calculation of damages based on environmental damage. The US Oil Protection Act created following the Exxon Valdez oil spill[10] in Prince William Sound, Alaska (now the second largest spill of oil in the US) was a US legal landmark for the calculation of damages. The Act allows for removal costs (for the clean-up of a spill) as distinct from 'damages' to natural resource property, subsistence use, revenue earning capacity and public services. Under the Act the National Oceanic and Atmospheric Administration (NOAA, the main federal marine science organization) is responsible for setting regulations to determine the assessment of the natural resource damage. For the Exxon Valdez a Blue Ribbon NOAA panel upheld the use of **Contingent Valuation (CV)** to assess damages and made recommendations on the interpretation and proper utilization of these studies in the legal context. The State of Alaska commissioned a CV study (Carson et al, 1992) which followed the NOAA recommendations for the use of CV methodology to calculate lost existence values from the spill. They estimated the existence value to the US population at $2.8 billion.[11] While the actual criminal and civil settlement with Federal and State governments amounted to $1.15 billion, the US Oil Protection Act enshrined the admissibility of such studies within the legal process. The final private civil payments from the Exxon Valdez oils spill are (25 years later) almost complete.

---

10  For full treatment of ecological, social and legal consequences *see* the 1700 entry bibliography of *Exxon Valdez* oil spill (Johnson and Rustin, 2013).

11  In order to protect themselves financially from the legal consequences of the spill, Exxon agreed a line of $4.8bn credit with the investment bank JP Morgan & Co in the face of a $5bn punitive damages bill. To mitigate the risk of non-payment the bank created the first modern Credit Default Swap in 1994(Gupta, 2012). A credit default swap is a financial contract whereby a buyer of corporate or sovereign debt in the form of bonds attempts to eliminate possible loss arising from default by the issuer of the bonds. This is achieved by the issuer of the bonds insuring the buyer's potential losses as part of the agreement. Credit default swaps were a major contributor to the global financial crash fourteen years later in 2008, with the financial ripples from the Exxon Valdez settlement contributing to the global financial collapse.

The damages to be paid by British Petroleum (BP) for the 2010 Deep-water Horizon oil spill in the Gulf of Mexico (the largest single spill in history) have yet to be awarded and the legal process is likely to go on for years, if not decades. Importantly, the US Congress has tasked a panel of experts (NRC, 2013) with evaluating the impacts of the oil spill on **ecosystem services** in the Gulf of Mexico. These concepts, applied properly, can produce defensible economic estimates of externalities caused by ecological damage and have the potential to send powerful financial signals into the social system.

These cases of compensation show how groups and institutions have been awarded compensation for environmental losses through judicial systems. Recently the main-streaming of ecosystem services concepts has resulted in growing attention to the idea of payment for ecosystem services. While such schemes are not yet widespread some successful examples do exist. For example, water from the Miyun reservoir in the Heibei province of China is used for agricultural production to flood-irrigate rice paddies in the municipalities of the province, and is also used to meet the domestic needs in the wealthier municipality of Beijing. Increasing conflict over water use has led to a series of collaborative programmes, one of which includes a rice to dry-land conversion scheme, where the Beijing municipality pay for the conversion of rice paddies to dry-land cultivation systems, thereby providing the provisioning service of higher water quality and quantities for use in Beijing and also providing the less well-off Heibei municipalities with a source of income (Zheng et al., 2013).

The diverse examples above illustrate important challenges to effective environmental management, as well as illustrating the direction of potential response options under the DPSWR. In both the cases of the IWC and the management of salmon in the Columbia River shifting human values, towards the conservation of whales in one case and genetic diversity within salmon in the other, have resulted in shifting management priorities based on social preference. The nutrient pollution examples of the Black and Baltic Seas illustrate the ecologically imposed challenges of improving environmental status under unpredictable systems as well as the challenges of addressing problems involving different nutrient sources spread widely over broad spatial scales.

In the oil spill cases and the payment for ecosystem services example there is a clear indication that the ecosystem service framework is beginning to meet its potential in quantifying externalities and eliminating market failures through incorporation into environmental decision-making. Our inability to completely restore ecosystems to their previous state means that the damage is already done and is not necessarily reversible by compensation. However, there is no doubt that such estimates can form a rational basis for the calculation of damages based on realities, and can act as a strong motivator for sectors to take environmental risks seriously.

Regulation is an essential component of Response and can be directed at any of the DPSWR elements, but regulation must also be designed pragmatically. The timescales of the legal process associated with the oil spill examples indicate the need for transparent, efficient and robust mechanisms to resolve natural resource management conflicts. Difficulties in enforcement and compliance, both internationally under the IWC and in the EU for the nitrates directive, suggest spatial mismatches between social institutions and ecological problems.

There are many potential Responses to a particular environmental problem, and the best response will depend on the nature of the social-ecological problem itself. In some cases Drivers are possible to eliminate while in others they are not, and in some cases a technological solution can break (or diminish) the link between Driver and Pressure. Where a species has been depleted, direct restocking or restoration may be possible, but our ability to control entire ecosystems and return them to previous states is limited. Targets and goals must also be achievable and adaptive and react to the changing realities of the ecosystem and the changing demands and values of social systems. Under such conditions, processes of **adaptive management** are required, allowing the manager to learn by doing. While the DPSWR can be helpful in indicating the broad potential options for response, there are many specific aspects of individual social and ecological systems that dictate how an effective response might be achieved.

## Suggested reading

Carson, R.T., Mitchell, R.C., Hanemann, W.M., Kopp, R.J., Presser, S., and Ruud, P.A. 1992. *A Contingent Valuation Study of Lost Passive Use Values Resulting From the Exxon Valdez Oil Spill*. A report to the Attorney General of the state of Alaska. 835pp.

NRC. 2013. *An Ecosystem services approach to assessing the impacts of the Deepwater Horizon oil spill in the Gulf of Mexico.* National Academies Press, Washington D.C. 235pp.

Zheng, H., Robinson, B.E., Liang, Y.C., Polasky, S., Ma, D.C., Ruckeshaus, M., Ouyang, Z.Y., and Daliy, G.C. 2013. Benefits, costs, and livelihood implications of a regional payment for ecosystem service program *PNAS 2013*; published ahead of print September 3, 2013, doi:10.1073/pnas.1312324110.

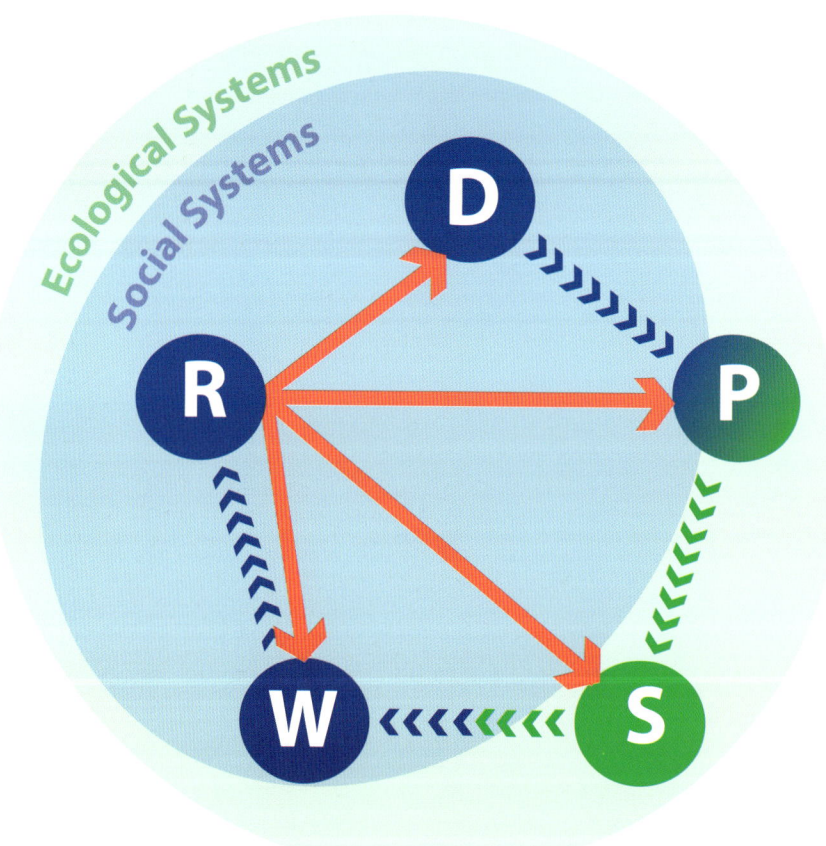

# 6  The Green Mountain

The previous chapters have covered stylized aspects of the DPSWR, giving diverse examples of the links between the DPSWR elements. This chapter is a short worked example of the use of the DPSWR to examine a single study site as a social-ecological system. It is based on a much more detailed integrated assessment that used the DPSWR approach (Al-Kalbani, 2015). This analysis illustrates the types of quantitative and qualitative data that needed to be combined to perform a complete DPSWR study. The chapter will examine how data were gathered, the choice of indicators and the area where assumptions had to be made, as well as problems with data availability and utility. Finally, the integrated picture developed through the DPSWR process will be used to assess the current approach to management in the area.

Al Jabal Al Akhdar (the Green Mountain) is a semi-arid mountain region at high altitude (1000–3075m above sea level) in the Sultanate of Oman. The region experiences rain and snow during winter and has hot, dry summers. The cool temperatures (relative to surrounding areas) and the availability of water have led, over centuries, to the evolution of an agro-pastoral oasis social-ecological system. Pomegranate trees and rose bushes for extraction of rose water are the main crops in the area. These are cultivated on irrigated terraces that run down the steep mountain sides (Figure 6.1). Goats, the main livestock graze the highland vegetation. The cool climate and verdant hillsides have made Al Jabal Al Akhdar a place of special significance for the Omani people. The construction, in 2006, of an asphalt road to connect the area to the rest of the country has facilitated the arrival of increasing numbers of tourists.

As an arid mountain, water is critical to the survival of the agro-eco-systems of Al Jabal Al Akhdar and the communities they support, and there has been a long history of water management in the region. There are three types of natural freshwater resources in the area: wadis, aflaj and

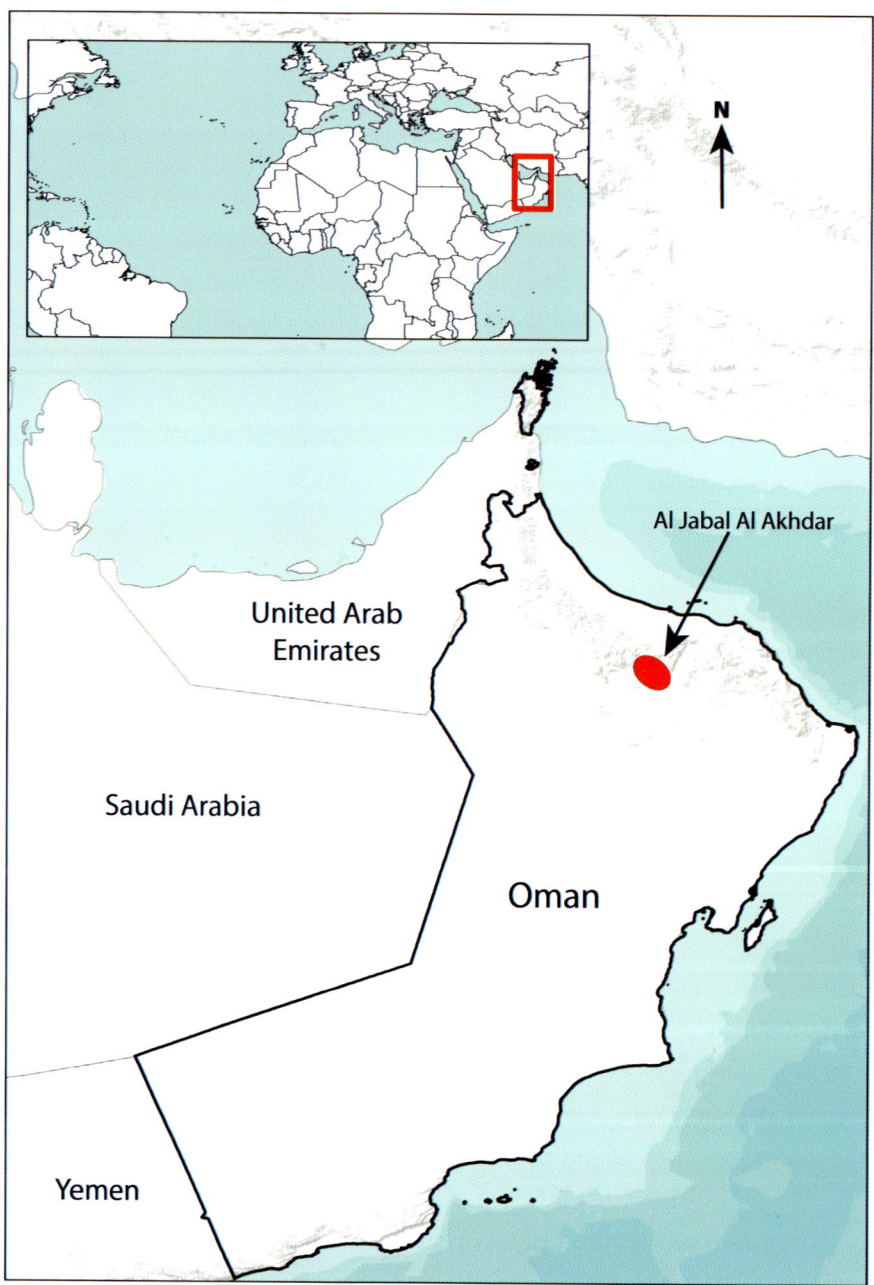

**Figure 6.1** Location of Al Jabal Al Akhdar.

groundwater. Wadis are dry, ephemeral river beds; Aflaj are an ancient technology for transport of water consisting of man-made channels, some of which have been in use for over a millennium (Al-Marshudi, 2001); both are now used to meet agricultural water needs. Groundwater, accessed via modern wells constructed by the Omani government, is currently the main source of drinking water in the area.

The major environmental problem in Al Jabal Al Akhdar in recent years has been a shortage of water. This water shortage is easily framed in the DPSWR terminology. Changing human activities, increasing numbers of people with increasingly affluent lifestyles (underlying Drivers) are placing greater demands on the water in the area, both for households and agriculture (immediate Drivers), resulting in increased water abstraction (Pressures). Increased water use has resulted in water shortages (State), which in turn has affected the lifestyles and practices of the people locally (Welfare). Some measures to increase the volumes of water available for use have been put in place and others are being developed (Response).

In order to get an initial overview of each of the DPSWR elements and to identify critical gaps in knowledge, the original study (Al-Kalbani, 2015) involved an extensive review of published and grey literature, government reports and studies. Subsequent fieldwork was used to eliminate some critical knowledge gaps. The state of water quality and quantity were assessed by collecting water samples for analysis over a one-year period as well as collecting data on the changes of groundwater levels and aflaj flow rates over the last few decades. A sophisticated multivariate statistical technique (cluster analysis) was used to choose representative sites for sampling and the samples were subjected to testing for a variety of water quality parameters. To examine the exogenous Pressures on the system, climate data, stretching as far back as the records for the area, were obtained from local meteorological stations. Further information on all of the DPSWR elements was gathered by designing, piloting, and eventually deploying surveys targeted at three groups: households, farmers and government officials/experts. The survey included questions on demographics as well as water use; opinions about the main Pressures on the ecosystem and perspectives on changes in water resource quantity area, as well as gathering information on the consequences of water shortage in people's daily lives. The main findings of the study, below, are broken out by the DPSWR categories.

## DRIVERS

According to government censuses, the population of Al Jabal Al Akhdar tripled between the years of 1970 and 2010 from less than 2000 to over 7000, (NCSI, 2012). The number of households has also grown along with population. There are now over 1600 housing units with only around 300 connected to wastewater treatment. The two wastewater treatment plants in the area are operating below capacity (MRMWR, 2014). The infrastructure in the area has expanded rapidly in the last few years, developments including a new road network, schools and the construction of groundwater wells and modern dams. Many residents have abandoned their small villages and moved to new modern housing in the area, fuelling a boom in construction. Offices, schools, mosques, hospitals, shops and restaurants have all been constructed and there are ambitious plans for further development of light industries including blacksmith's workshops, vehicle repair facilities and a car wash (Supreme Committee for Town Planning, 2011).

In tandem with the new development and the growing population and affluence in the area, traditional agricultural activities are continuing and are one of the main immediate Drivers of change. Pomegranates and roses, for extraction of rose water, are the main contributors to agricultural income (Agricultural Census, 2012–2013). Seventy per cent of local inhabitants work the land (Al-Riyami, 2006) and this figure increases during the growing and harvesting seasons for roses and pomegranates, when all members of a household generally share the seasonally increased burden of the work. These crops are grown on terraces irrigated using water from the wadis and aflaj. Two agricultural censuses have been conducted since the turn of the millennium (2004–2005 and 2012–2013). These censuses show that the area under cultivation has declined by 37% over the period (from 275 acres to 200 acres). If this trend were to continue, by 2050 less than 10% of the original cultivated area would survive.

Livestock husbandry is also an important aspect of agriculture in the area; goats in particular are raised for food and to generate income. Livestock numbers have been increasing and there has been a deterioration of the rangelands of Al Jabal Al Akhdar over the last decades. Overgrazing has resulted in reduced plant cover and dominance of inedible plants (Robinson et al., 2009). As an economic sector, agriculture contributes little to the overall Omani economy (about 3.7% of GDP); though it is the dominant consumer of water in Al Jabal Al Akhdar, reliable local figures for the economic benefits of agriculture are not available.

The recent development of Al Jabal Al Akhdar, the improved transport link of the asphalted access road and the construction of new hotels, in combination with the relatively mild climate and the spectacle of the traditional terraced agriculture, are attracting increasing numbers of tourists to the area. Most come to see the agricultural terraces and surrounding scenery (Al-Balushi et al., 2011). Over 130,000 people visit the area annually and the majority come in the summer months between May and October (Ministry of Tourism, 2014). The economic benefits created by a single hotel in Al Jabal Al Akhdar are estimated at US$390,000 annually (NCSI, 2014) and further development of the tourist industry is planned. During the tourist season there is high demand for water, which can result in the failure of supply to households as well as to tourist facilities, including the area's four hotels and the numerous holiday apartments and rest houses.

## PRESSURES

The combination of local Drivers, increasing affluence, growing population, agriculture and tourism result in increasing Pressure on the scarce water resources. Natural climatic variation also dictates the quantities of water available and changing climate acts as an **exogenous** Pressure on the system. For Oman, climate modelling suggests that future maximum and minimum temperatures are expected to increase by 1–2°C by 2040 (Al-Charaabi and Al-Yahyai, 2013) and analysis of local meteorological data for Al Jabal Al Akhdar indicates increasing trends in temperature (0.27°C per decade) since the late 1970s and declining trends in rainfall (9mm per decade) over the same period (Al-Kalbani et al., 2014). Figure 6.2 shows the five year moving average of rainfall illustrating this declining trend. The year of 1997 saw triple the average annual rainfall for the period and obscures an apparent downward trend in rainfall. During the period shown in the figure the Pressures on water placed by the Drivers of population, tourism and agriculture have either increased or continued.

Agricultural water abstraction places the largest Pressure on groundwater resources in the area. In Al Jabal Al Akhdar the agricultural terraces are irrigated by flooding, using waters from dams and aflaj, the levels of which

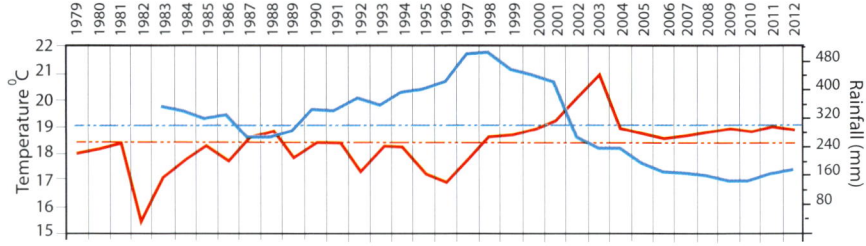

**Figure 6.2** Time series illustrating five-year moving average rainfall and mean annual temperature obtained from DGMAN (2014), Director General of Meteorology and Air Navigation, Public Authority of Civil Aviation, Muscat, Oman.

are dependent on recharge of water through rainfall. There is no measure of agricultural water consumption in the area, though in the general region it is thought to account for 92% of the total water consumed (MacDonald, 2013). The volume of water abstracted annually is likely to be decreasing as water becomes scarcer and the scale of agriculture is forced to shrink through lack of available water.

Household consumption of water from the two main groundwater wells in Al Jabal Al Akhdar has increased annually by 35% per year for over a decade, from 150,000m³ in 2001 to 580,000m³ in 2012 (PAEW, 2014). A significant proportion of this increase can be attributed to the burgeoning tourist industry. Though data are scarce, based on water tariffs paid by hotels and figures for the number of hotel visitors obtained from the local administration, the average daily water consumption per hotel visitor is between 641 and about 1300 litres, five to ten times the average daily domestic water consumption of individuals in the capital of Oman.

## STATE

The declining water supply through rainfall and the increased water demand are resulting in decreased availability of water. There are no accurate studies on the absolute volume of water in the Al Jabal Al Akhdar region. The one existing study of water balance, which has been carried out was at a larger spatial scale (MacDonald, 2013) suggests a deficit in water. That is, more water is abstracted in the area than is recharged through rainfall. While there is no direct measure for the amount of water in the area, one indicator of the State of the water resources is the depth to groundwater level. The depth from the surface to the level of groundwater is correlated with the amount of rainfall: the less rainfall, the deeper the level. Since 1997 the depth from the surface to groundwater has been continually increasing, indicating increasingly scarce water. Aflaj flow is also highly correlated with rainfall and the number of active aflaj is another indicator of water availability. The National Aflaj Inventory (MWR, 1999) conducted in the wet year 1997–1998 found only one of 72 local aflaj to be inactive, while in 2012–2013 only 38 aflaj in the region were active.

Water quality and water quantity are related aspects of the same problem. Studies of both irrigation water (reservoirs and aflajs) and drinking water (groundwater) quality found that most waters were fit for purpose according to Omani regulations. However, reservoirs and aflajs which were used for both drinking and irrigation in the past are now only fit for the purpose of irrigation, and there has been an unquantified loss in general water quality over the past decades.

## WELFARE

The declining State of the water resources and reduced area of cultivation (Figure 6.3) have resulted in considerable losses in agricultural income. The agricultural censuses indicate that the area of pomegranates, rose bushes, and nut trees have all declined while the deterioration of the pasture land has meant that farmers must now purchase goat meal rather than allowing goats to graze freely. Table 1 provides estimates of the economic losses to agriculture based on agricultural census estimates of the reduction in crop area (2005–2013), combined with interviews of locals on the productivity and economic yield of particular crops. The costs of buying goat meal due to the reduced quality of rangeland are also included.

**Table 1** Costs in US$ of the loss of agriculture and rangeland in Al Jabal Al Akhdar from 2005–2013.

| Product | Minimum | Mean | Maximum |
| --- | --- | --- | --- |
| Pomegranate | 2,593,626 | 3,242,033 | 3,890,439 |
| Nut Trees | 379,176 | 473,970 | 568,764 |
| Rose trees | 65,848 | 75,255 | 84,662 |
| Cost of goat meal | 17,467 | 26,200 | 34,934 |
| Total | 3,056,117 | 3,817,458 | 4,578,799 |
| Annual cost | 436,588 | 545,351 | 654,144 |

**FIGURE 6.3** Photographs showing the change in the agricultural terraces. **A)** summer 1997 (Al-Moharbi, K.A., 1997. Photograph of Agricultural terraces in Al Jabal Al Akhdar, Sultanate of Oman, taken in summer 1997.) **B)** summer 2014.

The water shortage results in economic costs to households that need to supplement their domestic supply. Currently households pay a charge of about $1.1 per metre$^3$ of domestic water and about US$0.66 million are spent by households on domestic water supply annually. Further costs are incurred when water shortages occur and householders are forced to purchase bottled water. Based on responses to the surveys (Al-Kalbani, 2015), the estimated average cost per household of purchasing bottled water annually is US$340. Considering the total number of households in the region, the total annual cost is about US$300,000. The combined annual costs of water shortage in the region (due to agricultural losses and reliance on bottled water) are of the order of US$1 million annually.

Considering the benefits created by a single hotel in Al Jabal Al Akhdar are about US$390,000 annually (NCSI, 2014), and that there are currently four hotels, the benefits generated by the tourism industry are likely of similar magnitude to the losses to households and to agriculture caused by reduced amounts of water. However, tourism is mainly based around the disappearing agro-ecosystems, which in turn are dependent on the supply of water. If tourism and agriculture are to co-exist there is a clear need to address the issue of water shortages.

## RESPONSE

The Omani government have recognized this trade-off and the major Response has been the construction of a pipe-line to pump desalinated water from the coastal region, up the mountains, to Al Jabal Al Akhdar. This Response acts on the State of the water resource, increasing water quantity directly by bringing new water to the region. The cost of establishing the project is approximately US$75 million. The costs of transporting the water from the coast to Al Jabal Al Akhdar (based on prices for a similar plant in Sana's City in Yemen) will be in the order of $3.6 per m$^3$, which would result in a doubling of the cost of potable water (when domestic and bottled are considered) to about US$2 million annually. Desalinated water will relieve some of the abstraction (Pressure) from the groundwater resources that supply the wells and aflaj, and ultimately the farms. The reduced amount of groundwater abstraction is not likely to halt the decline in agriculture since domestic use makes up only about 10% of the total. The costs of desalinated water would be prohibitive for agricultural use, given the relatively small economic value of the agriculture in the area. Logical justification for the introduction of desalinated water can therefore only be based around a considerably expanded tourist industry, which itself is reliant on the continuation of the agricultural practices.

The continued viability of the ancient agricultural systems is threatened by the Drivers and the exogenous Pressures causing declining water resource in Al Jabal Al Akhdar. The Response of pumping desalinated water to the area will offset the Pressure caused by increasing domestic water use (Driven by tourism) in the short term. However, using an energy-intensive technology to supply water also makes the area dependent on the supply of affordable energy, a factor which is outside local control. Future water supply in Al Jabal Al Akhdar will depend on global availability of fossil fuels and the geopolitics of energy supply in the long term. The additional carbon generated by the use of fossil fuels to pump the water will exert further Pressure on climate, thereby, in the long term, exacerbating the very problem responsible for the water shortages.

There are other potential Responses that may help to ensure the future of the Al Jabal Al Akhdar social-ecological system. Further development of the tourism sector requires consideration of the natural limitations of the ecological system; the appropriate scale for the Driver should be made explicit.

More efficient practices for both agricultural and domestic use of water could be adopted. For agriculture these could include further maintenance of aflaj, installation of modern irrigation systems, changing traditional agricultural practices, the use of greenhouses and the choice of alternative crops. For domestic and hotel water use there is potential for improved efficiency by expansion of the water distribution network, installation of conservation devices, meters, and the use of grey water. All of these measures would change the relationship between Driver and Pressure through technological means.

Additional infrastructure could also serve to increase the State of the water resource in the area for both agricultural and domestic use through construction of dams and more efficient rainwater harvesting and conservation systems.

The increased cost to households of domestic water supply are significant. In order to ensure the continued viability of households practising agriculture, government schemes to compensate local agricultural households for the elevated domestic water costs (acting directly on Welfare) may be in order.

There are no legal tools regulating or enforcing the installation of household water conservation devices in Oman. Introduction of such regulation and the inclusion of the cost of treating wastewater in the tariff for government water supply could help to internalize the problem of diminishing water supply, but these costs should be distributed appropriately. Community participation in water resources decisions may provide a basis for cooperation toward common goals. The people of Al Jabal Al Akhdar will need to develop a social system capable of adapting to decreases in rainfall and increases in drought.

As a case study in DPSWR the story of Al Jabal Al Akhdar illustrates some challenges of multi-disciplinary assessments. The data for the DPSWR elements of the Jabal Al Akhdar story came from diverse sources that had to be pieced together to form a complete picture. Much of the information about Drivers and Pressures came from relatively inaccessible grey literature and consultancy reports. Data were available at varying spatial scales and with differing reporting frequencies. Information held by individual government departments, which are not necessarily in regular communication, required considerable effort to gather and collate and required the development of a social network of cooperative individuals. Collecting the data for water quality and quantity analysis to understand the current State of the water resource required considerable time and physical effort to obtain. Sampling was in remote mountain locations and analysis of the samples required considerable capital and infrastructure.

All data come with their own limitations. Even the three-decade long meteorological time-series available for this study can only provide limited information on the long-term trends in climate. The natural variability in the data (for example, the rainy year of 1997) make it difficult to assess the effects of exogenous forcing. The absence of data on water storage in dams or flow rates of aflaj necessitated a reliance on imperfect indicators such as groundwater level and numbers of live aflaj. Gathering data to incorporate the human Welfare aspects included the development, testing and deployment of a survey instrument, and this required skills and techniques from the social sciences. The high cost of desalinated water would be prohibitive to its use for agricultures.

Even with perfect data the conclusions that can be drawn from a particular dataset are limited. Necessarily built on imperfect data, integrated assessments, such as this one, often face an additional challenge: understanding the ecological and social aspects of the system requires a multidisciplinary approach. An environmental scientist or ecologist may be comfortable, for example, with the statistical analysis of the variations in meteorological data, quantitative figures on means, minima and maxima, but is likely to be less comfortable handling the semi-quantitative information used to develop the estimates of economic value. Similarly, an economist or social scientist will lack the very specific sets of skills of a hard science background needed to analyse biological samples. There is no doubt that the integrated picture linking the State changes to human Welfare is key to identifying avenues for society to respond to the problem, but the holistic approach that this requires is not commonly taught in traditional environmental science or economic curricula.

For centuries the agro-pastoral oasis system of Al Jabal Al Akhdar has survived. Through history, this social-ecological system has been geographically isolated and relatively closed to the outside world. Historically, water availability has connected the social and ecological parts of the systems; the ability to store and manage water in the area has been developed over long periods of time and has dictated the bounds of agricultural production and the levels of human development. The Pressure of a changing climate is now placing further constraints on the practice of agriculture, demanding strategies for adaptation to ensure survival. The economic development in the area, the arrival of tourism, is placing additional demands and more competition for the scarce water resources on this already stretched ecosystem. Desalination and pumping can relieve the immediate State of drinking water quantity and meet the increased demand generated by tourism (though not agricultural needs), albeit dependent on global energy supply. Crucially, tourism is reliant on the agriculture, and the continued supply of agricultural water still remains a problem. Ensuring a continued supply of water for agriculture will require other social adaptations.

The story of Al Jabal Al Akhdar clearly illustrates the tension between a drive for development and long-term sustainability. The opening up of the transport links to the area that facilitate development have effectively removed a geographic barrier that allowed for small-scale sustainable management of the social ecological system. This system is in the process of adapting to the new, more globalized reality. Whether agriculture in Al Jabal Al Akhdar can survive in the long term will depend on an effective Response.

## Suggested Reading

Al-Kalbani, M.S., 2015. Integrated Environmental Assessment of Water Resources in Al Jabal Al Akhdar using the DPSIR Framework, Policy Analysis and Future Scenarios for Sustainable Management. PhD Thesis, University of Aberdeen, UK.

Al-Charaabi, Y., and Al-Yahyai, S., 2013. Projection of Future Changes in Rainfall and Temperature Patterns in Oman. *Journal of Earth Science and Climatic Change* 4 (5): 154–161.

Al-Marshudi, A. S., 2001. Traditional irrigated agriculture in Oman: Operation and management of the aflaj system. *Water International* 26 (2): 259–264.

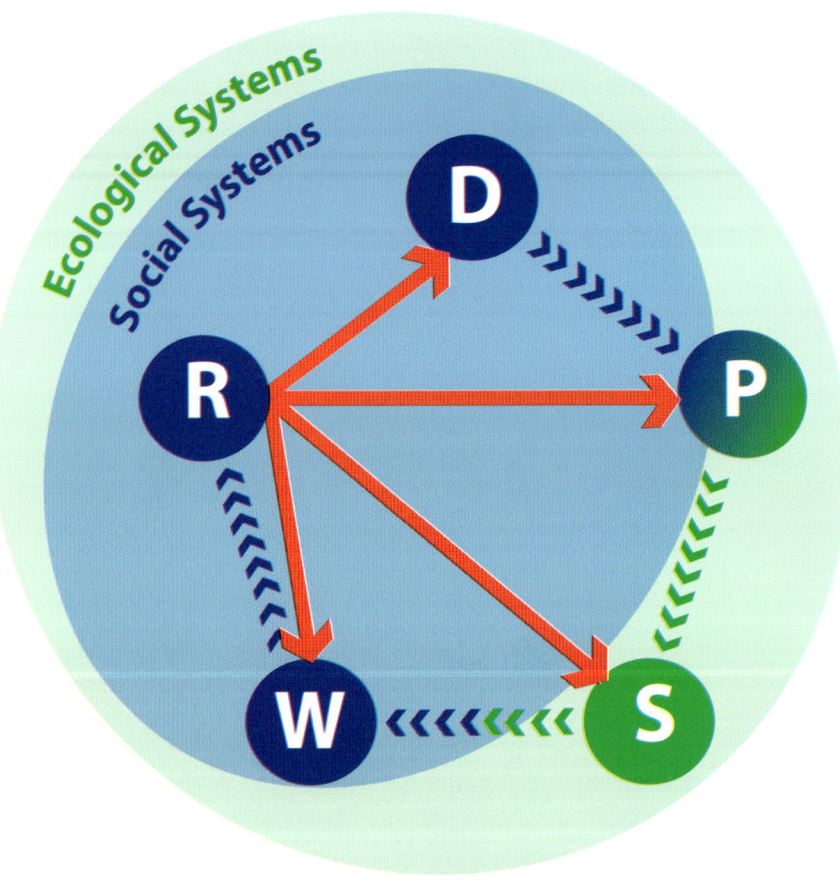

# 7 Balanced optimism

Human influences on the planet are now the dominant forces affecting the globe. This dominance can be observed in our abstraction of water (Postel et al., 1996), in global fluxes of essential nutrients (Vitousek et al., 1997), and in our oceans (Halpern et al., 2008). Society commandeers the fossil energy from the past to fuel the development of economies and to meet the wants and needs of an ever growing human population. Our impacts on the planet are now such that some claim we have entered a new era, the Anthropocene, an age when human activity is the dominant influence on the environment (Zalsiewicz et al., 2008). This unprecedented change in the relationship between Man and nature has happened quickly; humanity has gone from 100,000 years of development in a world that was constrained by natural resource availability at local scales to an era where natural resource constraints are global. The strategies we have evolved, as individuals and as societies, to survive in a non-resource limited planet need to be changed to meet this emerging reality. Adapting to this new situation is perhaps the single greatest challenge ever to face humanity. Natural resource constraints are factors that can drive societies to conflict and to war and dictate the patterns of future development; we can no longer afford to leave them at the fringes of political and economic thought.

In ecological systems many important processes occur at boundaries, from molecular transport across cell membranes that control the functions of life to the biogeochemical fluxes across the boundary of ocean and atmosphere that control global climate. Understanding the process at the interface between one system and another is essential to understanding how our planet functions. This is no less important at the interface between social and ecological systems.

To understand how societies can, or should, adapt to a crowded planet we need to understand the natural constraints posed by the ecological system (the limitations of resource use), their consequences for society, and the best ways to respond to the restrictions these constraints impose. Our responses will require social and economic changes as well as technological innovations. Ecological and economic concerns are fundamentally linked. Since human welfare is ultimately founded on the natural resource base provided by functioning ecosystems, both economics and ecology have essential roles in determining the future of human welfare. If we acknowledge this connection we must also recognize the need to balance the optimism of perpetual economic growth with the reality of ecological limitation. Though very different in their in their approach, ecology and economics are deeply related. Both are normative sciences: economics seeks to meet human welfare; ecology seeks to ensure the functioning of nature; these goals are interdependent and these two fields of expertise need to be brought closer together.

This need for transdisciplinary science, and the need for integration of the disciplines of economics and ecology in particular, has long been voiced (Daily and Ehrlich, 1999). The value of multidisciplinary research is becoming more widely recognized, but developing the social and intellectual capacity to put integrated approaches into practice remains a slow process. Society often favours excellence in a single field of expertise, and academic rewards systems, which measure success by publication, also favour specialization. This specialization can result in the development of academic silos, artificial distinctions between disciplines – for example, between environmental economists and ecological economists. These academic silos need to be abandoned.

Figure 7.1 is an abstract representation of this problem. In Figure 7.1a the central pillar represents a specific environmental problem with many different facets: social, ecological, and economic characteristics; the different patterns could represent, for example, different temporal frequencies. The complexity of the problem is not amenable to description or explanation by a single discipline. The rings (doughnuts, tori) depict individual disciplines, experts working within their academic silos; each discipline has the capacity to explain one facet of the problem. Figure 7.1b illustrates the situation where these individuals from all the disciplines come together to take an integrated approach. The connecting ring represents a community of practice of experts, sufficiently familiar with and knowledgeable of the expertise of the others to bring to bear a joint, multidisciplinary approach.

**Figure 7.1A and B** Abstract representation illustrating the importance of developing a transdisciplinary school of practice for environmental management (after Olsen et al., 2006).

The recent international interest in ecosystem services provides one example of the erosion of academic silos, bringing the communities of ecology and economics closer together in recognition of common goals. These disciplines also share a common etymological root, the Greek word *oikos,* meaning household. If economics studies the flows of income and expenditure of the household, ecology must manage the physical inputs and outputs. Now more than ever we need to recognize this inter-relationship and integrate these disciplines into the combined science of **oikology**.

The flexibility of the DPSWR and its value in organizing information from multiple disciplines and helping to frame problems in an accessible way is demonstrated by the case studies from around the world. Each case and each chapter has illustrated specific aspects of the Driver Pressure State Welfare Response, but the chapters have also served to illustrate major challenges facing us in the management of the environment.

The chequered history of the environmental movement is illustrated by the considerations of population in chapter 1. Here, early environmentalists used biological or ecological models to make unfounded predictions about future global catastrophes. They were too pessimistic; their failure was to ignore the adaptability and ingenuity of humanity. The ecological concept of carrying capacity was too simplistic when applied to humans. The famines predicted did not materialize, because humans did not reproduce according to immutable mathematical laws. Societies adapted technologically and socially to avoid the catastrophes predicted at the time of peak population growth. That is not to say that human societies should not be worried. Ecological footprints show that developed countries are using resources beyond the bounds of what the planet can provide in the long term. Solutions to the joint questions of optimal levels of population and affluence cannot be discerned with scientific methods; they involve value judgements. It is a question for global society whether population and affluence are maintained within limits through peaceful means, improved equity and voluntary moderation of consumption, or, as in the past, through brutality and global conflict. These questions are at the heart of global geo-politics. Economy, the environment, and politics are inseparable, and environmentalism must be incorporated into the mainstream of economic development. This is a long way from the status quo.

If we are to level the accusation of pessimism at the early environmental movement we must also acknowledge that the technological optimism and discounting of the future, so common in current economic practice, are dangerous. We need to foster societies that do not base their assessment of Welfare, as our present ones do, on short-term consideration of the levels of ever-growing consumption, but also on the recognition of the welfare generated by the ecological systems that surround us, and have enabled us to evolve and survive on the planet. We must measure and incorporate the value of ecosystem services more fully into our economic models, and understand the limits to growth imposed on us by our reliance on ecosystems.

As we struggle to implement policies to repair the damage done by past actions, and prevent further damage, it has become increasingly apparent that the ecosystems on which we depend for our well-being do not respond rapidly or predictably to our restoration efforts, but find alternative (and often less desirable) stable states. Just as the pessimism of the early environmentalists was short-sighted and underestimated the human capacity to adapt, so the historic approach to restoration of these ecosystems has been naïve, assuming that a reduction in Pressures would lead linearly to a return of historic state. Ecological systems (as well as social systems) are adaptive and unpredictable.

Managing common pool resources is notoriously difficult. The **'Tragedy of the Commons'** (Hardin, 1968) [1] summarizes the classic thinking on the subject. The argument goes that, where many individuals exploit a common pool resource, there is no incentive for anyone to conserve that resource because, by refraining from exploitation, the individual benefits less and leaves more to be exploited by others. That is, in a commons, conservation has no effect on the State of the resource and conservation actions are against individual self-interest. The classic Response to this problem has been the idea of transferring ownership right of commons to individuals or to governments[2]. Transferring ownership rights allows resources to be made excludable and more manageable (for example, the private ownership and stewardship of heather moorlands in Chapter 3). For many years management of common pool resource problems has been run on this basis.

The emerging science of ecosystem services is beginning to recognize the diversity of the properties of ecosystem services. These properties of rivalness and excludability of some ecosystem services (Chapter 4), which differentiate them from typical goods bought and sold on markets, offer potential for

1 The same man who argued in the 1970s for suspension of ethics under growing global population.
2 Personal preference for either individual or state ownership in such matters can neatly split individuals into political right and left; an individual may prefer a more social democratic government ownership or a liberal free market belief in individual ownership.

new ways of incorporating the values provided by ecosystems into systems of resource management.

Relatively recently the dogma of private or public ownership has been challenged. Empirical studies into ownership rights from multiple natural resource management systems have indicated that while some systems involving private or government ownership are successful (e.g. the heather moorland above), many are not (e.g. the disastrous quota system of the European Union Common Fisheries Policy) and successful resource management systems often follow specific **design principles** (Ostrom, 1990). Recognizing the different characteristics of ecosystem service supply, as well as the characteristics of successful management systems, offers the potential to design more appropriate institutions in the future.

The scale of environmental problems, the size of the globalized economic system, the mess of conflicting Drivers and competing value systems, and the interconnections between all these, can make the prospect of solving ecological problems seem impossible. There are many other books about the functioning of ecosystems or about the economically efficient extraction of natural resources. The aim of this book is to introduce a conceptual framework to help break down the complexity of social-ecological problems into their component parts, to organize our thoughts on these problems, and to make explicit the types of values and the trade-offs involved in solving these problems. Above all, this book offers some basic tools and concepts to help understand the links between ecosystems and the social systems they support, to balance the optimism of economics with the caution of ecology, and to introduce the rudiments of oikology.

### Suggested reading

Hardin, G. 1968. The Tragedy of the Commons. *Science* 162: 1243–1248.

Ostrom, E. 1990. *Governing the commons: the evolution of institutions for collective action*. Cambridge, Cambridge University Press.

# Glossary

**adaptive management**  an iterative process for continually improving management in the face of uncertainty. Learning by doing.

**biocapacity**  an elision of 'biological' and 'capacity', is the ability of an ecosystem to produce useful biological materials and to absorb carbon dioxide emissions.

**carrying capacity**  the number of individuals of a particular species that a given ecosystem can support without degradation of that system.

**compliance**  the extent to which individuals or economic sectors follow rule, obligation or regulation.

**command and control**  the direct regulation of an industry or activity by legislation that states what is permitted and what is illegal.

**common pool resources**  resources from which it is costly to exclude individuals due to their spatial characteristics.

**Contingent Valuation (CV)**  a survey-based technique used to elicit the individual's willingness to pay for a non-marketed resource.

**cultural service**  a non-material benefit obtained from ecosystems.

**demographic transition**  a shift from high birth and death rates to low birth and death rates occurring as countries become more economically developed.

**design principles**  a set of rules for successful management of common pool resources, which were formalized by Eleanor Ostrom (1990) as below:
1.  Define clear group boundaries.

2. Match rules governing use of common goods to local needs and conditions.

3. Ensure that those affected by the rules can participate in modifying the rules.

4. Make sure the rule-making rights of community members are respected by outside authorities.

5. Develop a system, carried out by community members, for monitoring members' behaviour.

6. Use graduated sanctions for rule violators.

7. Provide accessible, low-cost means for dispute resolution.

8. Build responsibility for governing the common resource in nested tiers from the lowest level up to the entire interconnected system.

**discount rate** weight given to future costs and benefits; the higher the discount rate the lower the value placed on benefits accruing in the future.

**Driver** an activity or process intended to enhance human welfare.

**Driver Pressure State Welfare Response (DPSWR)** a conceptual framework used to structure thinking, organize and categorize information about environmental problems.

## E

**Ecosystem Approach** a resource planning and management approach that integrates the connections between land, air, water, all living things, human beings and their activities and institutions.

**ecological footprint** a measurement of the amount of biologically productive land and sea area an individual, a region, all of humanity, or a human activity requires to produce the resources it consumes and absorb its carbon dioxide emissions. This measurement is compared to how much land and sea area is available.

**economic transition** a process whereby economies achieve a gradual reduction in ecological footprint.

**ecosystem service** an aspect of an ecosystem used directly or indirectly to yield human well-being.

**enforcement** the act of compelling compliance with a rule or obligation or regulation.

**excludable** a resource that it is possible to prevent individuals from using.

**Exogenous** something arising outside a particular system which is uncontrollable from the frame of the decision maker.

**externality** a cost imposed by Drivers or sectors in an economic market which is not reflected in that market.

## F

**final services** aspects of nature used directly to yield human well-being.

**fungible** mutually interchangeable.

**Future effect** the action of a current Pressure in a neighbouring ecosystem component, on future environmental State of the ecological system being managed.

## I

**Immediate Drivers** Immediate Drivers are activities that are proximal to at least one Pressure.

**input controls** regulations determining the amount of a specific Driver activity.

**intermediate services** aspects of nature used indirectly to yield human well-being.

## J

**joint production** the flow of many different benefits or services from a single process or ecosystem, or ecosystem component.

## M

**market failure** a situations where the allocation of goods and services by markets is not economically efficient (i.e. does not maximize production of goods and services, including ecosystem services).

**Memory effect** the action of a past Pressure on current environmental State.

**Multi-Criteria Analysis (MCA)** a technique used to evaluate options and make decisions that account for several criteria.

## O

**observed preference methods** techniques for measuring non-market values that are based on the real-life behaviours of individuals.

**oikology** the study of social ecological systems.

**opportunity cost** the benefit foregone by choosing one alternative action over another.

**output controls**  regulations determining the amount of Pressure or State change permissible.

**Pressure**  a means by which at least one Driver causes or contributes to a change in State.

**provisioning service**  material or energy outputs from ecosystems that provide benefits to humans.

**pure public goods**  goods where individual consumption leads to no loss of consumption by other individuals.

**R**

**regulating service**  components or processes of ecosystems that provide benefits to humans by regulating aspects of the environment, e.g. flood prevention.

**replacement cost**  estimate of the economic value of ecosystem services based on the costs of replacing these services.

**residence time**  the length of time that a particular material remains in a particular location.

**Response**  an initiative intended to reduce at least one Impact (State or Welfare change).

**restoration**  a Response to an environmental problem acting directly on the State of the ecosystem.

**rival**  property of a good or service whereby its use by one individual prevents use by another.

**S**

**spatial controls**  management measures, a subset of input controls that restrict the location of particular Driver activities.

**State**  an attribute or set of attributes of the natural environment that reflect its integrity as regards a specified issue (or change therein).

**stated preference methods**  methods of measuring non-market values based on the preferences expressed by responding to different types of economic survey instruments.

**stochastic**  something random or unpredictable.

**supporting service**  intermediate ecosystem services that make possible the provision of final services.

# T

**technological optimism** a belief that technology can find substitutes for resources and allow continued growth in populations and/or economies.

**temporal controls** management measures that restrict the timing of particular Driver activities.

**toll goods** goods that people can be prevented from using but those use by one person does not reduce the availability to others; for example, a park with an entry fee.

**Tragedy of the Commons** a situation arising where many individuals exploit a common pool resource. Often there is no incentive for anyone to conserve that resource because, by refraining from exploitation, the individual benefits less and leaves more to be exploited by others.

# U

**Underlying Driver** population, economic, social and technological factors that influence the level/nature of Immediate Drivers.

# W

**Welfare** a change in human welfare attributable to a change in State.

**WTP (willingness to pay) value** a hypothetical sum of money that an individual would pay for a non-marketed good or services.

# References

## Introduction

Balmford, A., Bruner, A., Cooper, P., Costanza, R., Farber, S., Green, R.E., Jenkins, M., Jefferiss, P., Jessamy, V., Madden, J., Munro, K., Myers, N., Naeem, S., Paavola, J., Rayment, M., Rosendo, S., Roughgarden, J., Trumper, K., and Teurner, R.K. 2002. Economic reasons for conserving wild nature. *Science* 297: 950–953.

Bonneville Power Administration, 2012. Annual Report Year in review and management's discussion and analysis. 52pp.

Bradshaw, C.J.A., Sodhi, N.S., and Brook, B.W. 2009. Tropical turmoil: a biodiversity tragedy in progress. *Frontiers in Ecology and Environment.* 7: 79–87.

Butler, V.L. and O'Connor, J.E. 2004. 9000 years of salmon fishing on the Columbia River, North America. *Quaternary Research* 62: 1–8.

Campbell, S.K. and Butler, V.L. 2010. Archaeological evidence for resilience of Pacific Northwest salmon populations and the socioecological system over the last ~7500 years. Ecology and Society 15 (1): 17. [online] URL: http://www.ecologyandsociety.org/vol15/iss1/art17/

Cooper, P. 2013. Socio-ecological accounting : DPSWR, a modified DPSIR framework, and its application to marine ecosystems. *Ecological Economics* 94: 106–115.

Coulthard, S., Johnson, D., and McGregor, J.A. 2011. Poverty Sustainability and human well being: a social well being approach to the global fisheries crisis. Global Environmental Change. 21: 453–463.

Craig, J.A. and Hacker, R.L. 1940. The History and Development of the Fisheries of the Columbia River. *Bulletin of the Bureau of Fisheries* 49: 133–216.

Ford, M.J. (ed.). 2011. Status review update for Pacific salmon and steelhead listed under the Endangered Species Act: Pacific Northwest. U.S. Dept. Commerce. NOAA Technical Memorandum NMFS-NWFSC-113, 281pp.

Geist, H.J. and Lambin, E.F. 2002. Proximate causes and underlying driving forces of tropical deforestation. *Bioscience* 52: 143–150.

Hamilton, L.C. and Butler, M.J. 2001. Outport adaptations: social indicators through Newfoundland's cod crisis. *Human Ecology Review* 8: 1–11.

Lackey, Robert T. 2000. Restoring wild salmon to the Pacific Northwest: chasing an illusion? In: Patricia Koss and Mike Katz (eds). *What We Don't Know about*

*Pacific Northwest Fish Runs – An Inquiry into Decision-Making.* Portland State University, Portland, Oregon, 91–143.

Landry, C.J. 2003. The wrong way to restore salmon. *PERC Reports* 21: 9–11.

Mantua, N.J. and Hare, S.R. 2002. The Pacific Decadal Osicillation. *Journal of Oceanography* 58: 35–144.

Mantua, N.J. and Hare, S. R., Zhang, Y., Wallace, J.M., and Francis, R.C. 1997. A Pacific interdecadal climate oscillation with impacts on salmon production. *Bulletin of the American Meteorological Society* 78: 1069–1079.

Naiman, R.J, Alldredge, J.R., Beauchamp, D.A, Bisson, P.A. Congleton, J., Henny, C.J, Huntly, N., Lamberso, R.N, Levings, C., Merrill, E.N., Pearcy, W.G., Rieman, B.E., Ruggerone,G.T., Scarnecchia, D., Smouse, P.E., and Wood, C.C. 2012. Developing a broader scientific foundation for river restoration: Columbia River food webs *PNAS* 109: 21201–21207.

Radtke, H.D. 1996. *The Cost of Doing Nothing: The Economic Burden of Salmon Declines in the Columbia River Basin.* Eugene, OR: Institute for Fisheries Resources, Northwest Regional Office.

Scholze, M., Knorr, W., Ameli, N.W., and Prentice, C.I. 2006. A climate-change risk analysis for world ecosystems. *Proceedings of the National Academy of Science* 103: 13116–13120.

US Federal Marine Mammal Protection Act 1972. http://www.nmfs.noaa.gov/pr/pdfs/laws/mmpa.pdf

Yaron, G. 2001. Forest, plantation crops or small-scale agriculture? An economic analysis of alternative land use options in the mount Cameroon Area. *Journal of Environmental Planning and Management* 44 (1): 85–108.

## Chapter 1

Assadourian, E. 2012. The path to degrowth in overdeveloped countries. In: State of the World 2012 – Moving toward a sustainable prosperity. Island Press.

Cohen, J.E. 1995. Population growth and Earth's human carrying capacity. *Science* 69: 341–346.

Commoner, B. 1972. A bulletin dialogue on the 'The Closing Circle': Response. *Bulletin of the Atomic Society* 28: 42–56.

Daly, H. 1973. *Toward a Steady State Economy.* W.H. Freeman and Co., San Francisco. 332pp.

Ehrlich, P.R. 1968. *The Population Bomb.* Ballating Books, New York. 223pp.

Ehrlich, P.R. and Holdren, J.P. 1971. Impact of Population Growth. *Science* 171: 1212–1217.

Evenson, R.E. and Golling, D. 2003. Assessing the Impact of the Green Revolution, 1960 to 2000. *Science* 300: 758–762.

Ewing, B., Moore, D., Goldfinger, S. Oursler, A., Reed, A., and Wackernagel, M. 2010. *The Ecological Footprint Atlas 2010.* Oakland: Global Footprint Network.

Hardin, G. 1974. Living on a lifeboat. *Bioscience* 24: 561–568.

Malthus, T. 1798. An essay on the principle of population. Electronic Scholarly

Publishing Project. http://www.esp.org.

Peng, X. 2011. China's demographic history and future challenges. *Science* 333: 581–587.

Pingali, P.L. 2012. Green Revolution: impacts, limits, and the path ahead. *Proceedings of the National Academy of Science*. 109: 12302–12308.

Young, C.C. 1998. Defining the range: the development of carrying capacity in management practice. *Journal of the History of Biology* 31: 61–83.

United Nations. 1948. Universal Declaration of Human Rights (adopted 10 December 1948 UNGA Res 217 A(III) (UDHR).

Wackernagel, M., Schulz, N.B., Deumling, D., Linares, A.C., Jenkins, M., Kapos, V., Monfreda, C., Loh, J., Myers, N., Norgaard, R., and Randers, J. 2002. Tracking the ecological overshoot of the human economy. *Proceedings of the National Academy of Science* 99: 9266–9271.

Waggoner, P.E. and Ausubel, J.H. 2002. A framework for sustainability science: a renovated IPAT identity. *Proceedings of the National Academy of Science* 99: 7860–7865.

Wang, C. 2012. History of the Chinese Family Planning program: 1970–2010. *Contraception* 85: 563–596.

## Chapter 2

Adeney, W.E. 1908. Effect of the new drainage on Dublin Harbour. Handbook to the Dublin District British Association. 378pp.

Barret, J., Johnstone, C., Harland, J., Van Neer, W., Ervynck, A., Makowiecki, D., Heirich, D., Hufthammer, A.K., Enghoff, I.B., Amundsen, C., Christiansen, J.S., Jones, A.K.G., Locker, A., Hamilton-Dyer, S., Jonsson, L., Lougas, L., Roberts, C., and Richards, M. 2008. Detecting the medieval cod trade: a new method and first results. *Journal of Archaeological Science* 35: 850–861.

Cloern, J.E. 2001. Our evolving conceptual model of the coastal eutrophication problem. *Marine Ecology Progress Series* 210: 223–253.

Enghoff, I.B. 1999. Fishing in the Baltic Region from the 5th century BC to the 16th century AD: evidence from fish bones. *Archaeofauna* 8: 41–85.

EU 2000 Directive 2000/60/EC of the European Parliament of the council of the 23rd October 2000 establishing a framework for community action in the field of water policy. Official Journal L 327 22/12/2000 1-73.

EU 2012 C-301/10: European Commision vs United Kingdom of Great Britain and Northern Ireland. (Failure of a Member State to fulfil obligations — Pollution and nuisance — Urban waste water treatment — Directive 91/271/EEC — Articles 3, 4 and 10 — Annex I(A) and (B)).

Finni, T. Laurila, S., and Laakkonen, S. 2001. The history of eutrophication in the sea area of Helsinki in the 20th century: long term analysis of plankton assemblages. *Ambio* 30: 264–271.

Mee, L.D. 2006. Reviving Dead Zones. *Scientific American* 295: 79–85.

Ménesguen, A., Perrot, T., and Dussauze, M. 2010. Ulva mass accumulation on

Brittany beaches: explanation and remedies deduced from models. *Mercator Ocean Quarterly Newselter* 38: 4–13.

Nausch, G. 2013. Water exchange between the Baltic sea and the North Sea, and conditions in the deep basins. *HELCOM Baltic Sea Environmental Fact Sheets*. Online. 14/7/14. http://www.helcom.fi/baltic-sea-trends/environment-fact-sheets

Neumann, T. 2007. The fate of river-borne nitrogen in the Baltic Sea – an example for the river Oder. *Estuarine Coastal and Shelf Science* 73: 1–7.

Oguz, T., Dippner, J.W., and Kaymaz, Z. 2006. Climatic regulation of the Black Sea hydro-meteorological and ecological properties at interannual-to-decadal timescales. *Journal of Marine Systems* 60: 235–254.

Oguz, T. and Velikova, V. 2010. Abrupt transition of the northwestern Black Sea shelf ecosystem from a eutrophic to an alternative pristine state. *Marine Ecology Progress Series* 405: 231–242.

O'Higgins, T.G., Cooper, P., Roth, E., Newton, A., Farmer, A., Goulding, I., and Tett, P. 2014. Temporal constraints on ecosystem management: definitions and examples from Europe's regional seas. *Ecology and Society* 19(4): 46.

Pilkshs, M., Kalejs, M., and Grauman, G. 1993. Influence of environmental conditions and spawning stock size on the year-class strength of eastern Baltic Cod *ICES Journal of Marine Science* 22: 1–13.

Ramankutty, N., Evan, A.T., Monfreda, C., and Foley, J.A. 2008. Farming the planet: 1. Geographic distribution of global agricultural lands in the year 2000. *Global Biogeochemical Cycles* 22: GB1003.http://dx.doi.org/10.1029/2007GB002952.

Röckstrom, J., Steffen, W., Noone, K., Persson, A., Chapin III, F.S., Lambin, E.F., Lenton, T.M., Scheffer, M., Folke, C., Schellnhuber, H.J., Nykvist, B., deWit, C.A, Hughes, T., vand de Leeuw, S., Rodhe, H., Sorlin, S., Snyder, P.K., Costanza, R., Svedin, U., Falkenmark, M., Karlber, L., Correll, R.W., Fabry, V.J., Hansen, J., Walker, N., Liverman, D., Richardson, K., Crutzen, P., and Foley, J.A. 2011. A safe operating space for humanity. *Nature* 476: 472–475.

United Nations – Department of Economic and Social Affairs. 2013. *World Economic and Social Survey*. 216pp.

Vahtera, E., Conley, D.J., Gunstafsson, B.G., Kuosa, H., Pitkanen, H., Savchuk, O.P., Tamminen, T., Viitasalo, M., Voss, M., Vasmund, N., and Wulff, F. 2007. Internal ecosystem feedbacks enhance nitrogen fixing cyanobacteria blooms and complicate management in the Baltic Sea. *Ambio* 36: 1–10.

Vitousek, P.M., Mooney, H.A., Lubchenco, J., and Melillo, J. 1997. Human Domination of the Earth's Ecosystems. *Science* 277: 494–499.

# Chapter 3

Anthony, A.W. 1924. Notes on the present status of the northern elephant seal, *Mirounga angustirostris. Journal of Mammalogy* 5: 145–152.

Bonnel, M.L. and Selander, R.K. 1974. Elephant seals: genetic variation and near extinction. *Science* 184: 908–909.

Connelly, N.A., O'Neill Jr., C.R., Knuth, B.A., and Brown, T.L. 2007. Economic impacts of zebra mussels on drinking water treatment and electric power generation facilities. *Journal of Environmental Management* 40: 105–112.

Davis, M.A., Chew, M.J., Hobbs, R.J., Lugo, A.E., Ewel, J.J., Vermeij, G.J., Brown, J.H., Rosenzweig, M.L., Gardener, M.R., Carroll, S.P., Thompson, K., Pickett, S.T.A., Stroberg, J.C., Tredici, P.D., Suding, K.N., Herenfeld, J.G., Grime, J.P., Mascaro, J., and Briggs, J.C. 2011. Don't judge species on their origins. *Nature* 474: 153–154.

Durie, A. 1998. Unconscious benefactors: grouse-shooting in Scotland, 1780–1914. *The International Journal of the History of Sport* 15: 57–73.

Elton, C.S. 1958. *The Ecology of Invasions by Animals and Plants*. Methuen, London.

EU 2000. Directive 2000/60/EC of the European Parliament and of the Council of 23 October 2000 establishing a framework for Community action in the field of water policy. OJ L 327, 22.12.2000.

Fraser Allander Institute. 2010. *An economic study of grouse moors*. Game and wildlife Conservation Trust, Scone, Scotland.

Grant, M.C., Mallord, J., Leigh, S., and Thompson, P.S. 2012. *The costs and benefits of grouse moor management to biodiversity and aspects of the wider environment: a review*. Royal Society for the Protection of Birds, Bedfordshire.

Harris, S., Morris, P., Wray, S., and Yalden, D. 1995. *A review of British mammals: population estimates and conservation status of British mammals other than cetaceans*. Joint Nature Conservation Committee, Peterborough.

Hooper, D.U., Chapin III, F.S., Ewel, J.J., Hector, A., Inchausti, P., Lavorel, S., Lawton, J.H., Lodge, D.M., Loreau, M., Naeem, S., Schmid, B., Setälä, H., Symstad, A.J., Vandermeer, J., and Wardle, D.A. 2005. *Effects of biodiversity on ecosystem functioning: a consensus of current knowledge. Ecological Monographs* 75: 3–35.

Huey, L.M. 1930. Past and present status of the northern elephant seal with a note on the Guadalupe fur seal. *Journal of Mammalogy* 11: 188–194.

International Court of Justice. 2014. Whaling in the Antarctic (Australia v. Japan: New Zealand Intervening), Judgment (Mar. 31, 2014), *available at* http://www.icj-cij.org/docket/files/148/18136.pdf

Knowler, D. 2008. Socio-economic pressures and impacts in BSC, 2008. In: Temel Oguz (ed.). *State of the Environment of the Black Sea (2001–2006/7)*. Publications of the Commission on the Protection of the Black Sea Against Pollution (Black Sea Commission) 2008–3, Istanbul, Turkey, 448pp.

Le Boeuf, Burney J. and Richard M. Laws (eds) 1994. *Elephant Seals: Population Ecology, Behavior, and Physiology.* University of California Press, Berkeley.

Lowe, S., Browne, M., Boudjelas, S., and De Pootrer, M. 2000. 100 of the world's worst invasive alien species: a selection from the global invasive species database. *Aliens* 12: 1–12.

Lyons, S.K., Smith, F.A., and Brown, J.H. 2004. Of mice, mastadons and men: human-mediated extinctions on four continents. *Evolutionary Ecology Research*

6: 339–358.

Millennium Ecosystem Assessment. 2005. *Ecosystems and Human Well-being: Synthesis*. Island Press, Washington, DC.

Mora, C., Tittensor, D.P., Adl, S., Simpson, A.G.B., and Worm, B. 2011. How many species are there on Earth and in the ocean? *PLoS Biol* 9: e1001127. doi:10.1371/journal.pbio.1001127. 137pp.

Nadafi, R., Blenckner, T., Eklöv, P., and Pettersson, K. 2011. Physical and chemical properties determine zebra mussel invasion success in lakes. *Hydrobiologia* 669: 227–236.Naderi, S., Rezaei, H.R., Pompanon, F., Blum, M.G.B., Negrinin, R., Nahash, H-R., Balkiz, O., Mshkour, M., Gaggiotti, O.E., Ajmone-Marsan, P., Kence, A., Vigne, J.D., and Taberlet, P. 2008. The goat domestication process inferred from large scale mitochondrial DNA analysis of wild and domestic individuals. *Proceedings of the National Academy of Science* 46: 17659–17664.

Redpath, S. and Thrigood, S. 1997. Red grouse and their predators. *Nature* 390: 547–547.

Shiganova, T.A., Dumont, H.J., Mikaelyan, A., Glazov, D.M., Bulgakova, Y.V., Musaeva, E.I., Sorokin, P.Y., Pautova, L.A., Mirzoyan, Z.A., and Studenikina, E.I. 2004. Interactions between the invading ctenophores *Menmiopsis leidyi* (A. Agassiz) and *Beroe ovata* Mayer 1912, and their influence on the pelagic ecosystem of the Northeastern Black Sea. In: Aquatic Invasions in the Black, Caspian and Mediterranean Seas. *NATO Science Series IV: Earth and Environmental Sciences* 35: 33–70.

Simoons, F.J. 1973. The sacred cow and constitution of India. *Ecology of food and nutrition* 2: 281–295.

Thornton, T. 1804. A sporting tour through northern parts of England and great part of the highlands of Scotland. (Printed for Vernor and Hood, 1804). Edward Arnold, London. 1896 edition, 332pp.

Thirgood, S., Redpath, S., Newton, I., and Hudson, P. 1995. Raptors and red grouse: conservation conflicts and management solutions. *Conservation Biology* 14: 95–104.

United Nations. 1992. Convention on Biodiversity. 1760 UNTS 79; 31 ILM 818 (1992).

Whitfield, P., McLeod, D.R.A., Watson, J., Fielding, A.H., and Haworth, P.F. 2003. The association of grouse moor in Scotland with the illegal use of poisons to control predators. *Biological Conservation* 114: 157–163.

# Chapter 4

Alcorn, S. and Solarz, B. 2006. The autistic economist. *Post-Autistic Economics Review* 38: 13–18.

Alexander, K., Janssen, R., Arciniegas, G., O'Higgins, T., Eikelboom, T., and Wilding, T. 2012. Interactive Marine Spatial Planning: siting tidal energy arrays around the Mull of Kintyre. *PloS One* http://dx.plos.org/10.1371/journal.pone.0030031.

Arciniegas, G., Janssen, R., Omtzigt, N. 2011. Map-based multicriteria analysis to support interactive land use allocation. *International Journal of Geographical Information Science* 25: 1931–1947.

Boyd, J. and S. Banzhaf. 2007. What are ecosystem services? The need for standardized environmental accounting units. *Ecological Economics* 63: 616–626.

Champ, P.A., Boyle, K.J., Brown, T.C. (eds). 2003. *A Primer on Nonmarket Valuation*. Kluwer Academic Publishers, Dordrecht, The Netherlands.

Chee, Y.E. 2004. An ecological perspective on the valuation of ecosystem services. *Biological Conservation* 120: 549–565.

Costanza, R. and Folke, K. 1997. Valuing ecosystem services with efficiency, fairness and sustainability as goals. In: Daly, G. (ed.). *Nature's Services: Societal Dependence on Natural Ecosystems*. Island Press, Washington DC, 49-70.

Costanza, R., de Groot, R., Sutton, P., van der Ploeg, S., Anderson, S.J., Kubiszewski, I., Farber, S., and Turner, R.K. 2014. Changes in the global value of ecosystem services. *Global Environmental Change* 26: 152–158.

Costanza, R., d'Arge, R., de Groot, R., Farber, S., Grasso, M., Hannon, B., Limburg, K., Naeem, S., Paruelo, J., Raskin, R.G., Sutton, P., and van den Belt, M. 1997. The value of the world's ecosystem services and natural capital. *Nature* 387: 253–60.

Daily, G.C. 1997. *Nature's services: societal dependence on natural ecosystems*. Island Press, Washington, D.C., USA.

Falk, A. and Szech, N. 2013. Morals and Markets. *Science* 340: 707–711.

Field, B.C. and Field, M.K. 2002. Environmental Economics: an Introduction. McGraw-Hill, New York, USA.

Fisher, B., Turner, K., and Morling, P. 2009. Defining and classifying ecosystem services for decision making. *Ecological Economics* 66: 643–653.

Gomez-Baggethun, E. and Ruiz Parez, M. 2011. Economic valuation and commodification of ecosystem services. *Progress in Physical Geography* 35: 613–628.

Jordan, S., O'Higgins, T., and Dittmar, J.A. 2012. Ecosystem Services of Coastal Habitats and Fisheries: Multiscale Ecological and Economic Models in Support of Ecosystem-Based Management. *Marine and Coastal Fisheries: Dynamics Management and Ecosystem Science* 4: 573–586.

Klain, S.C. and Chan, K.M.A. 2012. Navigating coastal values: participatory mapping of ecosystem services for spatial planning. *Ecological Economics* 82: 104–113.

Martin-Lopez, B., Montes, C., and Benayas, J. Economic valuation of biodiversity conservation: the meaning of numbers. *Conservation Biology* 22: 624–635.

Millennium Ecosystem Assessment (MEA). 2003. Ecosystems and human well-being: a framework for assessment. Island Press, Washington, D.C., USA.

Millennium Ecosystem Assessment (MEA). 2005. Ecosystems and Human Well-Being: synthesis. Island Press, Washington, D.C., USA..

O'Higgins, T.G., Ferraro, S.P., Dantin, D.D., Jordan, S.J., and Chintala, M.M. 2010. Habitat Scale Mapping of Fisheries Ecosystem Service Values in Estuaries. *Ecology and Society* 15: 7. [online] URL: http://www.ecologyandsociety.org/vol15/iss4/art7/

Sandler, R. 2012. Intrinsic Value, Ecology, and Conservation. *Nature Education Knowledge* 3 (10): 4.Schröter, M., van der Zanden, E.H., van Oudenhoven, A.P.E., Remme, R.P., Serna-Chavez, H.M., deGrooet, R.S., and Opdam, P. 2014. Ecosystem services as a contested concept: a synthesis of critique and counter-arguments. *Conservation Letters* (online) 1–10.

# Chapter 5

Carson, R.T., Mitchell, R.C., Hanemann, W.M., Kopp, R.J., Presser, S., and Ruud, P.A. 1992. *A Contingent Valuation Study of Lost Passive Use Values Resulting From the* Exxon Valdez *Oil Spill*. A report to the attorney general of the state of Alaska. 835pp.

Gupta, S. 2012. Credit default swap: regulations, changes and systemic risk. *Research Journal of Finance and Accounting* 3: 27–37.

Hardin, G. 1968. The Tragedy of the Commons. *Science* 162: 1243–1248.

HELCOM. 2009. Eutrophication in the Baltic Sea. An integrated thematic assessment of the effects of nutrient enrichment and eutrophication in the Baltic Sea region. *Baltic Sea Environment Proceedings* 115B. 148pp.

HELCOM. 2014. Eutrophication status of the Baltic Sea 2007–2011 – A concise thematic assessment. *Baltic Sea Environment Proceedings* 143. 40pp.

International Court of Justice. 2014. *Whaling in the Antarctic*. General List no. 148. March 2014.

ICPDR. 2007. Joint Action Program final implementation report. International Commission for the protection of the Danube river Vienna, Austria 214pp.

ICPDR. 2009. Danube River Basin District Management plan. Part A – basin wide overview. International Commission for the Protection of the Danube River Vienna, Austria 105pp.

ICPDR 2012. Interim report on the implementation of the joint program of measures in the DRBD. International Commission for the Protection of the Danube River, Vienna, Austria 41pp.

Jennings, S., Kaiser, M., and Reynolds, J.D. *Marine Fisheries Ecology*. Wiley-Blackwell, Oxford, UK. 432 pp.

Johnson, D.K. and Rustin, L.R. 2013. *Bibliography of* Exxon Valdez *oil spill publications*. Cambridge University Press, Cambridge, UK. 436pp.

Jota v Texavo, 1998. United State court of appeals.

Motohiro Kawashima. 2005. The Imagined Whale: How the Media Created a Sacrosanct Creature. *The Essex Graduate Journal of Sociology* (University of Essex, 2005). http://essex.ac.uk/sociology/research/publications/student_journals/pg/graduate_journal_vol5.aspx

Munkes, B. 2005. Eutrophication, phase shift, the delay and the potential return in the Greifswalder Bodden, Baltic Sea. *Aquatic Science* 67: 372–381.

NRC. 2013. *An Ecosystem services approach to assessing the impacts of the Deepwater Horizon oil spill in the Gulf of Mexico*. National Academies Press, Washington DC. 235pp.

Oguz, T., Akoglu, E., and Salihoglu, B. 2012. Current state of overfishing and its regional differences in the Black Sea. *Ocean and Coastal Management* 58: 47–56.

O'Higgins, T.G., Farmer, A., Daskalov, G., Knudsen, S., and Mee, L. 2014. Achieving good environmental status in the Black Sea: scale mismatches in environmental management. *Ecology and Society* (in press).

Paquet, P.J., Flagg, T., Appleby, A., Barr, J., Bankenship, L., Campton, D., Delarm, M., Evelyn, T., Fast, D., Gisalson, J., Kline, P., Maynard, D., Mobrand, L., Nandor, G., Seode, P., and Smith, S. 2011. Hatcheries, Conservation,and Sustainable Fisheries – Achieving Multiple Goals: Results of the Hatchery Scientific Review Group's Columbia River Basin Review. *Fisheries* 36: 547–561.

Reider, S. and Wasserstrom, R. 2013. Undermining democratic capacity: myth-making and oil development in Amazonian Ecuador. *Ethics in Science and Environmental Politics* 13: 39–47.

Tonnessen, J.N. and Johnsen, A.O. 1982. *The History of Modern Whaling*. C. Hurst and Company, London.

Waples, R.S. 1999. Dispelling some myths about hatcheries. *Fisheries* 24: 12–21.

Williamson, K.S., Murdoch, A.R., Pearsons, T.D., Ward, E.J., and Ford, M.J. 2010. Factors influencing the relative fitness of hatchery and wild spring Chinook salmon (*Oncorhynchus tsahwytscha*) in the Wentachee River, Washington, USA. *Canadian Journal of Fisheries and Aquaculture Science* 67: 1840–1851.

Zheng, H., Robinson, B.E., Liang, Y.C., Polasky, S., Ma, D.C., Ruckeshaus, M., Ouyang, Z.Y., and Daliy, G.C. 2013. Benefits, costs, and livelihood implications of a regional payment for ecosystem service program *PNAS* 2013; published ahead of print September 3, 2013, doi:10.1073/pnas.1312324110

## Chapter 6

Agricultural Census (2012/2013), Department of Statistics, Directorate General of Planning and Development, Ministry of Agriculture and Fisheries, Muscat, Oman.

Agricultural Census (2004/2005), Department of Statistics, Directorate General of Planning and Investment Promoting, Ministry of Agriculture and Fisheries, Muscat, Oman.

Al-Bulushi, A.S., Al-Mukhtar, B., Al-Hayis, A.J., Orfan, M., Hamza, J., Al Busaidi, Y.S., and Al Busaidi, N.H. 2011. *Socioeconomic Study of Tourism Development in Al Jabal Al Akhdar*. A Study conducted by Sultan Qaboos University in collaboration with Ministry of Tourism, Oman, in Arabic.

Al-Charaabi, Y. and Al-Yahyai, S. 2013. Projection of Future Changes in Rainfall and Temperature Patterns in Oman. *Journal of Earth Science and Climate Change* 4 (5): 154–161.

Al-Kalbani, Mohammed Saif. 2015. Integrated Environmental Assessment of Water Resources in Al Jabal Al Akhdar using the DPSIR Framework, Policy Analysis and Future Scenarios for Sustainable Management. PhD Thesis, University of Aberdeen, United Kingdom, in progress.

Al-Kalbani, M.S., Price, M.F., Abahussain, A.A., Ahmed, M., and O'Higgins, T. 2014. Vulnerability Assessment of Environmental and Climate Change Impacts on Water Resources in Al Jabal Al Akhdar, Sultanate of Oman. *Water* 6: 3118–3135.

Al-Marshudi, A.S. 2001. Traditional irrigated agriculture in Oman: operation and management of the aflaj system. *Water International* 26 (2): 259–264.

Al-Riyami, Y. 2006. *Agriculture Development in Al Jabal Al Akhdar*. Symposium on Economic Development of Al Jabal Al Akhdar Area, Oman Chamber of Commerce and Industry, Nizwa, Oman, 11 September.

DGMAN. 2014. Director General of Meteorology and Air Navigation, Public Authority of Civil Aviation, Muscat, Oman.

MacDonald, M. 2013. Water Balance Computation for the Sultanate of Oman, Final Report submitted to the Ministry of Regional Municipality and Water Resources, Muscat, Oman.

Ministry of Tourism. 2014. Data on Number of Tourists and Tourism Projects in Al Jabal Al Akhdar. Department of Statistics and Information, Muscat, Oman. Unpublished data.

MRMEWR. 2001. National Aflaj Inventory, Summary Report. Ministry of Regional Municipality, Environment and Water Resources, Muscat, Oman.

MRMWR. 2014. Ministry of Regional Municipalities and Water Resources. Data on wastewater treatment plants and sewer networks, unpublished data, Directorate General of Regional Municipality and Water Resources, Nizwa, Sultanate of Oman.

NCSI. 2012. National Centre of Statistic and Information. Census 2010: Final Results, General Census on Population, Housing & Establishments 2010, Muscat, Sultanate of Oman. NCSI. 2014. Data on hotels. Unpublished data. National Centre for Statistics and Information, Muscat, Sultanate of Oman.

PAEW. 2014. Data on water abstraction and consumption from wells in Al Jabal Al Akhdar. Public Authority of Electricity and Water, Al Jabal Al Akhdar Office, Nizwa, Oman. Unpublished data.

Robinson, M.D., Al-Harthi, L.S., Al- Nabhani, S. and Al-Busaidi, B. 2009. Livestock Diets and Range Conditions on the Saiq plateau, Oman. In: Victor, R. and Robinson, M. D. (eds.). *Al Jabal Al Akhdar Initiative. Mountains of the World: Ecology, Conservation and Sustainable Development. Proceedings of the International Conference, Sultan Qaboos University, Sultanate of Oman, February 10–14, 2008.*

Supreme Committee for Town Planning. 2011. Al Jabal Al Akhdar Area Development Master Plan. In Arabic.

# Chapter 7

Daily, G.C. and Ehrlich, P.R. Managing Earth's Ecosystems: and interdisciplinary challenge. *Ecosystems* 2: 277–280.

Halpern, B.S., Walbridge, S., Selkoe, K.A., Kappel, C.V., Micheli, F., D'Agrosa, C., Bruno, J.F., Casey, K.S., Ebert, C., Fox, H.E., Fujita, R., Heinemann, D., Lenihan, H.S., Madin, E.M.P., Perry, M.T., Selig, E.R., Spalding, M., Steneck, R., and Watson, R. 2008. A global map of human impact on marine ecosystems. *Science* 319: 948–952.

Hardin, G. 1968. The Tragedy of the Commons. *Science* 162: 1243–1248.

Olsen, S.B., Sutinen, J.G., Juda, L., Hennessey, T.M., and Grigalunas, T.A. 2006. *A Handbook on the Governance and Socioeconomics of Large Marine Ecosystems*. Coastal Resources Centre, University of Rhode Island. 94pp.

Ostrom, E. 1990. *Governing the Commons: the Evolution of Institutions for Collective Action*. Cambridge University Press, Cambridge, UK.

Postel, S.L., Daily, G.C., and Ehrlich, P.R. 1996. Human appropriation of renewable fresh water. *Science* 271: 785–788.

Vitousek, P.M., Mooney, H.A., Lubchenco, J., and Melillo, J. 1997. Human Domination of the Earth's Ecosystems. *Science* 277: 494–499.

Zalasiewicz, M.W., Smith, A., Barry, T.L., Coe, A.L., Brown, P.R., Brechley, P., Cantrill, D., Gale, A., Gibbard, P., Gregory, F.J., Hounslow, M.W., Kerr, A.C., Pearson, P., Knox, R., Powell, J., Waters, C., Marshall, J., Oates, M., Rawson, P., and Stone, P. 2008. Are we now living in the Anthropocene? *Geological Society of America Today* 18: 4–8.

# Index

Page numbers in *italics* denote figures